LEAPFROGGING THROUGH THE UNIVERSE

KENNETH E. BURCHETT

LEAPFROGGING THROUGH *The* UNIVERSE

AMITY AMERICA PRESS
Branson

Amity America Press
Branson, Missouri

ISBN: 978-1-7350442-5-5
Library of Congress Control Number: 2022915925

On the cover: Great Spiral Galaxy M81 (NGC 3031).
Photo © R. Jay GaBany, Cosmotography.com.

In Memory of Katie Griffin

Albrecht Dürer. *Melencolia I.* 1514.
Engraving, 9 5/8 x 7 9/16 in. (24.5 x 19.2 cm.)
Gift of R. Horace Gallatin.
National Gallery of Art.

Contents

**The Orion Nebula.
A place where stars are born.
Photographed by the Spitzer Space Telescope.
NASA/JPL-Caltech.**

PREFACE

In our perception of the solar system and the universe, we accept the fact that enormous objects lie suspended in harmonious patterns guided by natural forces of immense power. We extrapolate from these facts a crude knowledge of the greater cosmos. Yet, the reality of the universe remains incomprehensible. The mind does not grasp events of such magnitude. No one has ever experienced the true sensory physicality of Earth and the Sun, let alone the galaxy and beyond. Such experiences remain indescribable by human perception.

How then does one experience the universe? The expectation that we are a part of a finite, teleological process, by which we may describe the physical universe, becomes illogical in the presence of its seemingly infinite scale.

The act of trying to see beyond the boundaries of our present reality is a unique human action, which in the end has no practical consequence except the ability of the act itself to enlarge awareness of human uniqueness. We exercise our being when we try to figure out what and how the universe is. To imagine the cosmos and all it contains is to feel the expansion of the inner self by our own creativity, and through the imagination and creativity of others before us. Like distant horizons that once enchanted seafaring adventurers, the opening of the universe captivates the human spirit and releases humanity from the provincialism of small realities.

This book is a story about the parting of the veil of the universe, tales of discovery, and the many frontiers of space that silently beckoned previous generations. It describes the mysteries of the known universe and reveals them in ways that not only explain but also inspire. It is a story of astronomy told in a blend of history, art, science, philosophy, mythology, and insight, which takes the reader from our own planet to the edge of the unknown. It is about people, discoveries, ideas, and the connecting events that make the universe what humankind thinks it is.

Galaxy Messier 81 in the Constellation Ursa Major, 11.8 million light years from Earth. © R Jay GaBany. Used with permission.

I

INTRODUCTION

1

The Algonquin people believed — as did many Indian nations — that a great being imbued all things with an unseen supernatural force. The Algonquians named this force Manitou. All Native American cultures related to the Algonquin people shared similar ideas. Variations of the Manitou belief existed throughout the tribes of the Eastern Woodlands, among the Blackfeet and Cheyenne of the Plains, and in the Salish and neighboring tribes of the Far West. In these societies, all natural objects contained souls the same as people. Friendly spirits became sacred guardians, and all spirits sacred and evil had the power to intervene in people's lives. The force of Manitou found its source in the stars, Moon, and Sun. Like the beliefs of many of the surviving religions of the world, the Great Soul was in all things.

From the beginning of humanity, people have looked up at the stars in the evening sky to wonder at the magnificence of what they saw, and to imagine within the brilliance of the stars a separate existence in that place. It is human nature to want to understand being, and when not understood, to accept as a condition of faith that which is envisioned amid the mysteries of the unknown.

2

Over the centuries, the veil of the mystery of the previously unknown universe has fallen away; what once was magical no longer is. Yet we are no closer to an understanding of the purpose of human existence than in the beginning. The facts of the universe are as incomprehensible as the dogmas of spiritual belief. This common inability to grasp these two great ideas, one represented by astronomy and the other by religion, continues to draw the mind and soul toward the boundless and unfathomable domain of space that is at once heaven and the cosmos.

There are many different representations of the universal Great Soul. In early America, the Northwest Indians celebrated reverence for sacred things by carving images on totem poles. Those of the Southwest made secret rock paintings. Indians in the East developed sophisticated language codes to express spiritual meaning. Cultures all over the world left their own unique symbols for embracing the unknown.

The use of language as a tool of expression is particularly interesting. Words often divine the universe. As a case in point, the Ojibwa Indians, an Algonquin-related culture of the Eastern Woodland group surviving today mainly in northern Canada have in their language a hierarchy of the Algonquin Manitou. Articles of their language distinguish vital things from inanimate ones, things that are alive from those that are not. In a code similar to how other languages use articles to distinguish male and female forms, in French, for example, the Ojibwa separate the spirit from the non-spirit. In the Ojibwa language, the stars receive the article describing the presence of life, while the body, the house of the soul, has inanimate reference. The body lives only through its internal spirit while the universe is alive and synonymous with the Ultimate Soul. The sacred works of the world's other great religions take no less of a point of view. As the discoveries of the universe unfold, the question of omnipotent being is frequently magnified and rarely diminished.

3

From the earliest time of human presence forward, belief in a magical kingdom is a common finding in all cultures. No kingdom was more magical than the cosmos. Reverence for the universe as a magical place runs diffused throughout every part of human history, eloquently expressed in colorful mythologies, and enduringly preserved in the rich artifacts of the archaeological past. Manitou gave ultimate meaning to the lives of the American Indians and shaped the cultures of the North American continent. Throughout the world, similar beliefs sprung from the earliest people to foreshadow a modern intrigue with the stars. Religion fulfilled a timeless and universal interest in association between the stars and a common heaven.

Modern fascination with the vast known universe is undiminished from what it was long ago. The language used to describe the universe may be different today from before, its frontiers more defined, and the celebrations of its grandeur less ostentatious than that of the tribes of the distant past. Nevertheless, the attraction of the universe as a magical kingdom is unchanged. Knowledge about the universe, rather than diminish its mystery, has instead inspired the further search to uncover in the stars the allegorical if not the true spirit of human existence.

Beauvais Manufactory. after Guy-Louis Vernansal. *Les Astronomes,* from *L'Histoire de l'empereur de la Chine*. ca. 1697–1705. Wool and silk, 125 ½ × 167 in. (318.8 × 424.2 cm.) The J. Paul Getty Museum, Los Angeles.

THE PORTIERE OF PISCES

The Sun, with all the planets revolving around it, and depending on it, can still ripen a bunch of grapes as though it had nothing else in the Universe to do.

Galileo Galilei.
Attributed, 1564-1642.

Etienne Delaune. *Astronomy*. 1540-83. Engraving, 2 ¾ × 2 1/16 in. (7 × 5.3 cm.)
The Metropolitan Museum of Art.

II

THE OBSERVATORY

4

In the late 16th century, the Danish astronomer Tycho Brahe shattered the ancient belief in an unchangeable universe when he proved that a supernova he observed was a distant star. A supernova is the name given to an exploding star. Until Tycho's time, scholars believed the universe to be immutable. Change was impossible. An explosion in the universe meant that change in the universe was not only possible but a reality.

A century and a half after Tycho's discovery, Edmond Halley, using the ancient Greek catalog of Hipparchus, revealed that some of the stars charted by Hipparchus had moved. These two discoveries—by Tycho and Halley—showed that the stars not only experience change, they are also in motion, moving through the universe.

1.1. Tycho Brahe, Danish astronomer, ca. 1590. Museum of Fine Arts, Houston.

5

Before the era of the telescope, Tycho Brahe was the greatest astronomer of all time. Because he was one of the most famous figures in 16th century Europe, a number of portraits exist of him (**1.1**). A sometimes controversial and eccentric figure, he wore a false nose after suffering mutilation in a duel. Most likenesses editorialize his disfigurement while others show the facial reconstruction that he made of gold, silver, and wax.

Soon after Tycho's observation of the supernova, the King of Denmark built him an observatory on the Danish island of Hven (present Ven) in 1576 as a reward for the young astronomer's ability at an early age. There Tycho spent the next twenty years of his life completing some of the most important work ever done in astronomy. His development of unique astronomical instruments along with a demanding insistence on detail made possible new star charts with a degree of accuracy previously unknown.

History recorded observatories as early as 300 B.C.E., in Greek Alexandria. They probably existed in some form much earlier in Mesopotamia and Egypt. However, the modern observatory dates only to the sixteen hundreds and the development of the telescope. Tycho's observatory (**1.2**) was short lived, as were many such facilities prior to the invention of the telescope, and it closed in 1597. Nevertheless, his model proved to be the template for future observatories throughout Europe and elsewhere.

1.2. Tycho Brahe's Stjärneborg observatory on Ven, 1595. ***Atlas Major*****, Amsterdam, 1662.**

In 1599, Tycho moved to Prague to become the imperial mathematician, joined there by his assistant, a young Johannes Kepler. When Tycho died suddenly in 1601, Kepler inherited his notebooks. Tycho's carefully described and notated observations and measurements of the stars and planets enabled Kepler to calculate the orbits of the planets and set the stage for the discoveries of Sir Isaac Newton who, in turn, uncovered laws of mechanics and material behavior, which three hundred years later enabled space exploration.

6

Time and space in the universe

The idea that the universe is a changing and dynamic place, as Tycho discovered, was difficult for people then to accept after centuries of theological belief to the contrary. Resistance to new ideas about the universe was not, however, unique to the past. So persistent was the concept of a stable, unchanging universe among some scientists into the early 20th century that Einstein felt compelled to include a constant as part of his theory of relativity. Modern science was not prepared to accept the breadth of Einstein's claims about a mutable universe. The cosmological constant eventually became obsolete when subsequent discoveries about the universe proved Einstein correct.

When Einstein first introduced his theory, there were those who could not accept the idea that the universe was in motion to the extent that Einstein believed it to be. It took years for the scientific community to digest his predictions. Ironically, subsequent revelations about the high velocities of objects at the edge of the universe gave even Einstein pause to rethink some of his own theories about the changing universe.

7

No one knows how large the universe is because its furthermost limits remain unseen, to know even if it is limited or if it is infinite. Some indication of its scale, however, comes from knowledge of the stars and galaxies that make up this vast space. There are between a million and 10 trillion stars in a galaxy, depending on the galaxy's size; up to 10,000 galaxies in a cluster of galaxies; and numerous clusters of galaxies combined

1.3. At the edge of the universe, revealed by the Webb Telescope. NASA, ESA, CSA, and STScI NASA.

into super clusters. There may even be clusters of super clusters (**1.3**)! From this, it is easy to see how predictions of the number of stars in the known universe exceed numbers in the zillions, an indeterminate number.

8

Astronomers estimate the age of the universe to be between 14 and 20 billion years old. Some observers place its beginning more conservatively at 11.5 to 13 billion years, some even less at below 10 billion. The broad consensus at one time was for around 18 billion years. However, over time, this decreased, pared down to about 14 billion years, give or take a billion years, which is all to say that the age of the universe is unknown.

Such predictions of a finite age for the universe come mostly from adherents of the open, expanding universe theory. According to this theory, the calculation of a date for the beginning of time derives from working backwards from the most distant cosmic objects. Since light requires billions of years to reach Earth from the outer reaches of the universe, and since light came into existence more or less with the beginning of the universe, the universe must be as old as its most-distant object is. The most-distant object yet detected is predicted to be about 14 billion light years away. Figure **1.3** is a small section of the deepest-ever view of the universe taken by NASA's James Webb Space Telescope, which reached to

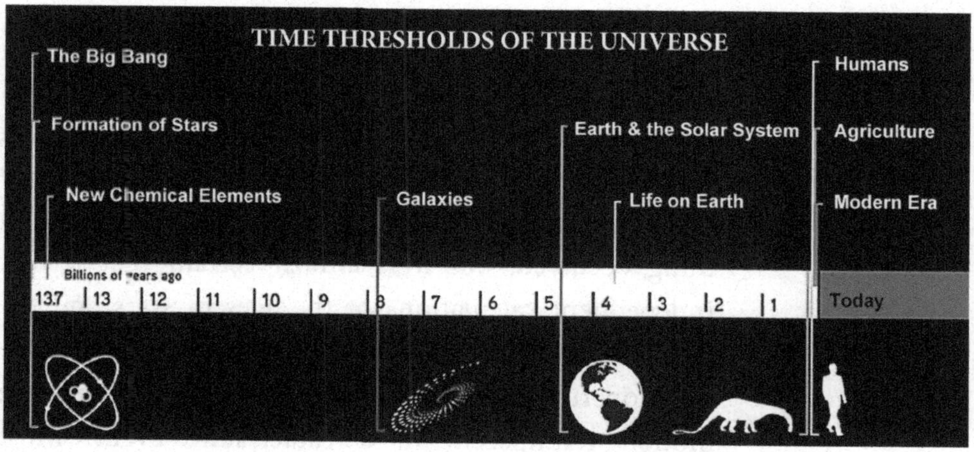

1.4. Historical and astronomical time, a time line of the formation of the universe.

the visible horizon of the cosmos. The furthermost object observed was approximately 14 billion light years distant. Therefore, this assumes the universe is at least 14 billion years old.

From the time of the beginning of the universe, cosmic events are marked in millions of years. The stars move and the constellations change but only noticeably over the course of a thousand human generations (**1.4**). This is why for centuries the ancients thought the stars were immovable and unchanging. The history of all civilization will not record its thousandth human generation for yet another 15,000 years. Therefore, historical time can account for only an infinitesimal fraction of astronomical time. On the astronomical calendar, five thousand years of human history is nothing. Time and space references in world events have little relationship to cosmic descriptions. It is necessary to enlarge greatly the line of human history in order even to see it on the astronomical time line. Compared to the history of the universe compressed into a century of time, a representation of all the recorded deeds of humankind would be less than one second. The clock of earthly existence is not the clock of the cosmos. On the cosmic time line, the history of human existence is infinitesimal.

9

The Big Bang

Modern astronomers believe that the universe began with an explosion called the Big Bang, a name that conjures up a vision of a massive explosion sending energy and particles out in a spherical pattern in all directions at once. No one will ever know how such a thing occurred, but it is almost certain that it did. Independent calculations of the ages of stars and the expansion of the galaxies away from each other all have the properties of bodies caught up in a cosmic explosive motion reminiscent of a cataclysmic event. Most predictions about the universe rest on the Big Bang theory.

10

A common description of the universe is the totality of existence of matter and energy and the space that contains it. Existence in the universe manifests in a space-time continuum that marks the transfiguration of matter to energy and vice versa.

In the beginning, there was no space, and there was no time. At the instant of creation, the motion of energy defined both time and space. As energy accrued into matter, time and space evolved to become a definite, if incomprehensible, duality. As the universe rapidly grew to unimaginable proportions, gravity became the unifying, constructive force. Stars were born — and are still being born — of the gas and dust spewed out of the cosmos. In many regions of the universe, stars are being born in the same process that marked the beginning of time. Deep inside columns of molecular hydrogen and dust, new stars are appearing. Gravity contracts the dense, cool gas into lumps that ignite under their own pressure. Eventually the dust-clouds around the stars evaporate, and the emerging stars become radiant points

1.5. The creation of stars in the Eagle Nebula. Hubble Heritage Team, NASA, ESA.

of light in the sky. The universe comprises billions and billions of these stellar embryonic centers. The pictured example (**1.5**) is located in the Eagle Nebula, and is more than 25 trillion kilometers (15.5 trillion miles) in size. Stars are forming deep inside the towering structures.

11

Companions of darkness and light

The future of the universe may lie not with the stars but with their dark invisible companions. Standing in a dark room, one will not see the burners on a hot stove until one of them begins to glow. The other burners are there but unseen, except perhaps by the reflected light of the one luminous burner. Thus, it is with things in space. Meanwhile, light is only one kind of radiated energy. There are many objects in space that do not radiate light, some seen only by sophisticated electronic means, and some that remain undetected at all by any means. The sky is full of things that are not visible.

Nevertheless, just as the kitchen range must reach a high enough temperature to emit visible energy, an object like a star must do likewise. Not all units of mass reach such high temperatures. Some that do cannot sustain them forever. A few celestial bodies move in and out of visible range, while others accumulate energy levels so high and masses of matter so great that they pass beyond the visible spectrum forever, never to reenter it. For every unit of mass that is radiating light, there may be as many as 10 dark units, a number variously estimated by astronomers to be as low as three units or as high as 16. Many dark bodies are likely to be different in form from anything previously known.

The history of civilization records that there is in the brightness of the cosmos a timeless, universal mystique. There is mystery also in its darkness. Yet we cannot describe what we cannot see, except by its effects and perhaps through the eye of the imagination. Therefore, the visible and not the invisible primarily confirm the reality of existence. The history of the universe is writ large in light. Therefore, while the darkness of the universe and its invisible energy has opened dramatic new chapters in the record of astronomy, the story of light is still the more interesting tale in the ever-unfolding saga of the universe. Some of our most profound experiences with light originate with the deeper universe.

12

Starlight

The physical size and age of the universe are beyond human understanding. Therefore, its story is one of trying to reduce the unknown, and perhaps the unknowable, to a scale and to a process that speaks to the human ethos. Because light reveals knowledge, light is and has always been at least with respect to the universe a kind of portal into eternity, a way of expanding our existence on Earth to larger proportions. The cosmos permits this; indeed, it invites an expansion of the soul through the stars and through a host of other visual phenomena.

Each point of light in the sky has a characteristic brightness that is its signature of individuality, an individuality that further promotes our identity with these distant and unknown companions of time and space. Moreover, when individual stars come together as

1.6. Trail of the Milky Way. The Milky Way arcs above American Indian ruins in Hovenweep National Monument, Colorado-Utah. National Parks Service.

one glowing mass, we find our hope for an organized, systematic universe enticingly rewarded. The Milky Way (**1.6**), for example, is a glowing trail of organized stars that indicates the shape of our own galaxy.

1.7. Earthrise seen from the Moon. Apollo 8 Crew, NASA.

Whether from the points of light that are the stars and galaxies that fill the seemingly infinite expanse of space to the nearby star we call the Sun, whose radiant mass enables all life as we know it, the light of the universe is omnipresent in timeless, reassuring cycles of brightness.

13

Events on Earth are the most common sources of our experience with light. This is due to our dependency on artificial lighting and the reflective surfaces of the natural environment. Earth, Moon, and planets are our most familiar celestial experiences; indeed, they are our only celestial experiences with natural light reflection, except for occasional comets. The crew of Apollo 8 took a famous photograph (**1.7**) that shows the reflective surface of Earth from the distant perspective of the Moon. Astronauts took this picture of Earth rising over the Moon in 1968.

The universe is littered with reflective objects like Earth and Moon, but we can see very few of them. The Moon and a few planets are visible because they are nearby, and the reflection of the Sun makes them visible. They are not self-luminous, but they are close enough for the eye to see them by their reflected light. They are the only non-self-luminous objects out of all that exist in the universe that are visible by eye from Earth.

14

The Sun

The Sun reveals our existence and makes life possible. It has risen and set for millions of millennia, and it will for millions to come. We see the Sun, feel its heat, and rely on its radiation to provide not just the joy of seeing, but the knowledge gained from the actual revelations of sunlight, and the metabolic necessities of an ever growing, healthy existence.

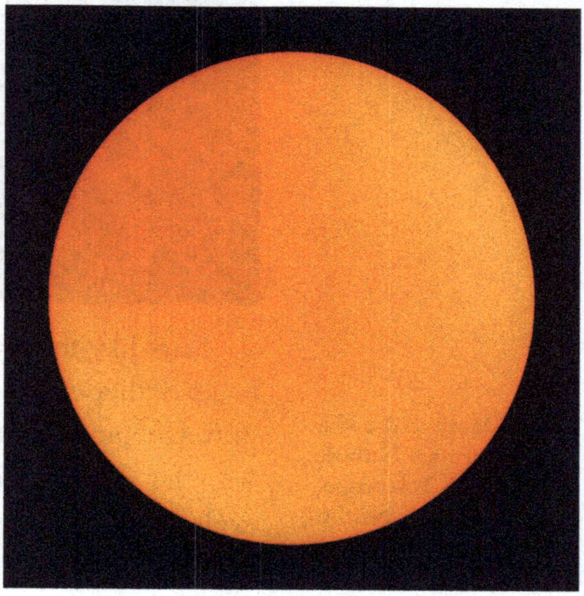

1.8. The face of the Sun unmarked by sunspots.

The Sun is one star that we cannot safely look at, at least not without special eye protection. Likewise, the brightness of the Sun renders impossible normal photography of its disk, except at dawn and dusk when the atmosphere shields its light, or when obscured by haze and fog. Without eye protection, the Sun is too luminous to be viewed when only a few degrees above the horizon without inflicting permanent damage to the eye, a hint of its unearthly power. The Sun appears small when seen from Earth. Yet the size of Earth relative to the Sun is as a period on this page compared to the image (**1.8**) in the photograph, to give some idea of their comparative scale. The significance of this relationship magnifies in the uncountable billions of stars that exist in the universe, many with their own tiny earth-like dots.

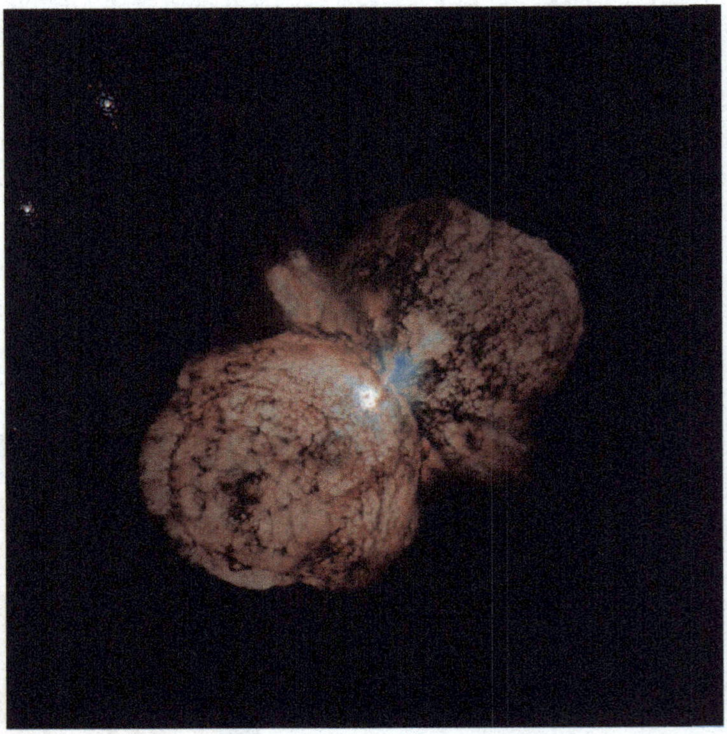

1.9. Death of the star Eta Carinae. Hubble Space Telescope, NASA.

Stars like the Sun are born as giant nuclear energy engines; they evolve and die. A picture of a region of the universe known to be a birthplace of stars appeared in **1.5**; the death of a star, meanwhile, we see in the explosive force of the giant star Eta Carinae (**1.9**). Debris from this explosion is moving away from the core of the star in the directions of its poles at the speed of 2.4 million kilometers (1.5 million miles) per hour. Despite its great self-destructive violence, Eta Carinae has been for many years one of the brightest stars in the sky, a phenomenon that continues to puzzle scientists. At the rate at which it appears to be disintegrating, it should have faded long ago.

The average life span of a star measures in billions of years. The life span of the Sun, for example, is predicted to be between nine and 11 billion years. No one knows for sure exactly when the Sun was born or how long it will continue to shine, but it will shine at least until it burns up its supply of fuel. If, as many believe, Earth is four to five billion years old beginning when the solar system coalesced around the Sun, and if the life span of the Sun is 10 billion years, then predicated on the idea of a sustainable system, the Sun may last at least another five to six billion years. At the end of its life when it does die it will enlarge like Eta Carinae to engulf the planets, then shrink to 1 percent of its current size, becoming a white-hot dot glowing in the night sky like a dying ember, until it fades slowly away to itself become a companion of the darkness of the universe.

Tycho Brahe proved that the universe is a living and perishable place. Succeeding generations unlocked just how awesome it really is. Unlike anything experienced on Earth, nothing that Tycho or anyone could have imagined matches the grandeur of what lies beyond the planet Earth.

Johann Sadeler I, after Maerten de Vos. *Geometria*, from *The Seven Liberal Arts*. ca. 1570–1600. Engraving and etching, 5 13/16 × 4 5/16 in. (14.7 × 11 cm.) The Elisha Whittelsey Collection, The Elisha Whittelsey Fund, 1949. The Metropolitan Museum of Art.

ENERGY AND THE CREATION OF FORM

We had the sky up there, all speckled with stars, and we used to lay on our backs and look up at them, and discuss about whether they was made or only just happened.

Mark Twain.
Adventures of Huckleberry Finn, 1884.

Giovanni Domenico Tiepolo. *Creation of the Universe.* 1727-1804.
Pen and gray ink with brown wash on laid paper, 6 $^{13}/_{16}$ x 9 $^{3}/_{8}$ in. (17.3 x 23.8 cm.)
National Gallery of Art.

III

COALESCENCE

15

Existence is the perpetual balance between matter and energy. In this respect, the most complete representation of existence is then the universe. Humankind understands intuitively that the universe contains in finite portions all that exists, has ever existed, or will ever exist.

All material existence begins as free energy and ends up as free energy. The metaphorical life cycle embodies this principle whether in the physical laws of natural existence or in the religious precepts of a spiritual transformation that governs it. Being comes from the unknown and returns to the unknown. The ultimate source of existence — the modern cosmic Manitou — is free energy passing through the universe without material presence.

A star by its fiery radiance signals a place in the universe where energy is no longer free but bound by material presence in a distant and fixed field of time and space. Between the stars are the great voids of space, the conduits of energy that convey the mysteries of the universe. Stars fill the evening sky, each one a distant beginning, its energy traveling freely through time and space, and carrying with it a record of its material past. At any place in the universe where there is light there is, or once was, an object, something shining and visible, sending out a message of its existence, its energy expanding through the universe to remind all who see it of the timeless process of creation.

It is by the stars, and not the darkness, that we know the universe. Figure **2.1** is a view of a tiny piece of the

2.1. Alborz Mountains and evening sky in moonlight. Photo © Babak Tafreshi.

evening sky and the objects visible within it, seen in a way that has defined from the beginning the ordinary understanding of the universe. A first-quarter Moon illuminates the landscape. The large star on the left is Sirius, the brightest star in Earth's sky. Three stars align in the belt of the Constellation Orion. There is no suggestion in this photograph of the many billions of stars that are beyond the visual field, nor of the extraordinary kingdom to which they belong. Before the telescope, this is how the universe appeared, distant, tranquil, and majestic.

16

Visible and invisible energy

We call energy emitted from a star radiant energy, or radiation. Visible radiant energy, or light, reveals to us the vastness of space because its manifestation is in the brightness of celestial objects. Brightness is that primordial quality by which we know the stars and galaxies. It is the visual interpretation of radiant energy. Radiant energy gives a thing its luminous quality and causes it to be bright enough to see. Human visual perception translates luminous radiant energy into brightness. Light is visual radiant energy; luminance is the optical measure of light; and brightness is the visual interpretation of luminance. Differences in the brightness of cosmic objects seen on Earth begin with different amounts of visible radiant energy. This visible energy and the variations in its luminosity and perceived brightness tell us a great deal about the universe and the mysterious behavior of the material in it.

There are many metaphors of spirituality connected to luminous energy because light is synonymous with reality. Invisibility, on the other hand, although potentially more powerful than the visible, retains the stigma of being a mysterious unknown force for which there is no generally understood reality.

Not all energy is visible. Only the energy called light is, and it amounts to only a tiny fraction of all the energy that exists. Visible energy is to invisible energy as the tip is to an iceberg. What we see as light is small when compared to the energy that lies out of sight, moving invisibly through the universe. The stars and galaxies are just the visible markers of a mostly unseen and unknown place.

17

Invisible energy is everywhere, not just in outer space. In the example of a heating element of an electric kitchen range, heat radiates from the range element, yet the heated burner appears no different from one that is cold because heat is invisible. However, as the hot element reaches its highest setting, it will soon begin to turn red and glow. The energy is now visible as well as felt. Some of the radiant energy reached the visible light stage, and the burner became luminous. If it is very luminous then vision perceives it to be bright; low luminosity means it is dim or less bright. Such is the case with stars as well; the hotter ones are brighter, the cooler ones are less bright.

Another well-known common example of invisible radiation is the TV remote control, which everyone knows can send invisible signals across an empty space. Invisible energy travels over large distances to create a picture on a TV set, or sound on a radio. We do not see or feel this radiation. The ability to send and receive invisible energy signals across open space, undetected by human perception, is one of the great discoveries of modern science. Modern astronomy has made it possible to use invisible energy to *see* the invisible universe.

18

Applications to appliances and home entertainment are just some of the many ways invisible energy affects our lives. For most people, the awareness of invisible energy is limited to a few common types, like heat and other electrical and fuel-generated examples that come from living on Earth. However, many forms of invisible energy saturate the universe, some of them different from anything experienced on Earth. For instance, cosmic rays that arrive unseen from deep space are normally deadly if not blocked by Earth's atmosphere. Ultraviolet radiation, the energy of the Sun most responsible for a Sunburn, cannot penetrate the upper atmosphere in amounts sufficient to cause widespread health problems. The atmosphere blocks potentially harmful energy from outer space. All natural radiation on Earth in normal amounts is safe. The natural environment of our planet protects us uniquely.

19

Energy and matter

A basic premise of existence is that energy is invincible, neither created nor destroyed; it can only change and transform. Energy becomes matter, and matter becomes energy. Matter and energy are two different versions of the same thing. Architect Louis Kahn put it this way, "All material in nature, the mountains and the streams and the air and we, are made of Light which has been spent, and this crumpled mass called material casts a shadow, and the shadow belongs to Light."

Kahn's poetic view mirrors the reality of the universe. In the ebb and flow of cosmic events, energy forms into stars and stars combine with other stars to become star clusters and galaxies. Sometimes stars and galaxies explode, destroying matter and releasing energy to begin the creation cycle all over again.

2.2. Albert Einstein seated at his desk in the Patent Office at Bern, Switzerland. When Einstein graduated from college in 1900, he was unable to find a job in the academic profession. With only average grades to show, he was without steady employment for two years. It was during his time that he published his revolutionary ideas on the relationship of energy, matter, time, and space. Photo by F. Hoyle.

20

The inseparable relationship between matter and energy was first described in 1905 by Albert Einstein in his famous equation $E=MC^2$. Figure **2.2** pictures Einstein about the time he published his famous discovery. He was 26 years old when he wrote five brief papers on the behavior of light and the structure of time and space. These five papers, written in a four-month period, changed the world. His paper on energy and matter, the source of his famous equation, introduced the atomic age and forever changed the understanding of the universe.

Einstein was an unlikely success story. He failed his initial entrance examination to the Swiss Federal Polytechnical School (ETH). When he finally did gain entrance and subsequently graduated, his mentors wrote him off as unemployable. It is said that one of his teachers, when asked by the young Einstein's father what profession his son should study, replied, "It doesn't matter; he'll never make a success of anything." In fact, young Einstein appears to have been very good as a patent clerk. After acquiring his doctorate in 1905 from the University of Zurich, he took a job at the Swiss Patent Office, where he stayed until 1909 when he moved to Zurich to become an assistant professor of physics at the University. He was obliged to remain a patent clerk for four years after his landmark publications while the scientific community tried to digest his revolutionary ideas.

Einstein remained relatively unknown to the outside world until 1912, and the world did not fully appreciate his work until much later. Einstein moved to the United States in 1932, settling at Princeton, New Jersey, where he lived and worked until his death in 1955. From an obscure beginning, Einstein went on in his lifetime to publish more than 300 scientific papers and books, win a Nobel Prize, and become the most celebrated man in modern science. Einstein later would say that although the writing of his famous paper on general relativity took only five days, the idea had been with him since age 16.

21

The equation $E=MC^2$ means energy is equal to mass, or matter, multiplied times the speed of light, times the speed of light again. This produces an incredibly large number. Each relative metric unit of mass is equal to 90 billion relative units of energy. There is enormous potential explosive energy in a very small amount of material. Enough energy is contained in one kilogram (2.2 pounds) of uranium to destroy a large city or two. The amount of uranium needed to make an atomic bomb would fit inside a coffee cup. Einstein's discovery was the basis of the atomic bomb.

22

Since energy and mass are variations of the same thing, energy is equivalent to matter freed to accelerate to great velocity; conversely, matter is arrested energy. The Sun, for example, is a giant package of arrested energy (**2.3**). Its material constantly transforms into free energy in a steady, sustained stream of visible and invisible radiation, some of it reaching Earth, some not. Eruptions of the Sun's surface may produce giant irregularities that extend far out into space, in some instances, as much as 500,000 kilometers (300,000 miles), and farther in distance than from Earth to the Moon. Solar energy from some of these flares is strong enough to reach Earth. A telescopic view of the Sun's

2.3. The Sun's energy. Solar and Heliospheric Observatory.

active thermal surface shows energy patterns of erupting prominences on its surface, active gasses radiating in the Sun's corona that depict both gravitational and magnetic energy forces, and the churning thermal energy active on the Sun's surface.

The first atomic bomb had an explosive energy force equivalent to 20,000 tons of TNT (**2.4**). The largest nuclear bomb ever tested was a 45-60 million-ton hydrogen device — 3,000 times as powerful as the first atomic bomb — detonated by the Soviet Union in the 1960s. On a scale of cosmic proportions, imagine the mass of the Sun converted to energy and released all at once, as in an atomic bomb. The Sun's explosive potential is equivalent to 900 trillion (nine followed by 26 zeros) of the largest nuclear bomb ever detonated on Earth.

2.4. Atomic energy. The first atomic bomb pictured with its mushroom cloud rising in the pre-dawn hours above the desert near Alamogordo, New Mexico, in 1945.

Fortunately, the conversion of matter to energy is not always necessarily cataclysmic. The Sun's release of energy occurs over a long period of time and in gradual amounts, not like an atomic bomb where a comparatively tiny piece of matter such as uranium instantaneously — within a few billionths of a second — releases all of its energy in one gigantic burst of destructive force. If the Sun were to explode like that, the explosion would obliterate the entire solar system.

23

Mass and gravity

The definition of physical existence in time and space needs three qualities. These qualities are energy, matter, and mass. Matter is any physical thing that occupies space; energy alone does not occupy space. Mass, meanwhile, takes up space but has the additional quality of being a unified body. Mass is always matter but matter may not always be mass. Nevertheless, in practice the two terms, namely, matter and mass, usually are interchangeable.

Mass, however, unlike matter, further expresses the outcome of a consolidation process whereby matter and energy conform to a specific volume such as would occur when a significant amount of gravity is present. From a reversed perspective of this idea, gravity is like a sponge whose shape manifests in the absorption of matter about a cosmic point to make something persist as a unified structure, that is, to create a mass. Each cosmic body exists defined in time and space by a center of mass and shaped by the force of gravity.

Another way to look at a center of mass is as a center of gravity. In the vacuum of space, the evolution of mass and the construction of gravity is the process by which the universe transforms from energy to form. In this process, according to Einstein's theory, a unit of energy is a particle that has gravitational attraction but no mass. When the tinniest of these particles attract the tinniest unit of matter, form begins to grow as a point of gravity. More and more units of matter are attracted to each other to create a coherent, sustainable bit of material to become a mass, a unified thing in space. This can happen very rapidly. As energy builds around a gravity point, material accumulates faster and faster as it arranges and rearranges itself uniformly about its center. As the mass of the object grows, the force of gravity grows with it in a kind of self-evolving process of creation, a process limited only by the amount of attractable material within reach of the gravitational field of the object. Energy accelerates under the pressure of condensing matter. Moreover, the more gravity the more matter will be attracted, and the more matter attracted, the greater the mass and its gravitational field, until the accelerated energy of the condensing mass reaches a height of activity that can no longer be contained and energy bursts forth into space in a continuous stream of

radiation. This is how stars are born. Figures **2.5** and **2.6** illustrate one of many ominous regions of the universe churning with matter and energy where stars are in the process of forming.

2.5. The Horse Head Nebula. This distinctive shape of cosmic gas appears as an amorphous cloud of energy and cosmic material in a mysterious region of the universe where stars are forming.

As a star releases, or radiates, energy, the mass that produced the energy continues to rearrange itself, settling into a more compact whole. Matter packs more tightly, and the sphere of its gravitational mass draws closer around its center. As a result, gravitational force increases even more. Because mass tends to compact relative to its loss of radiated energy, gravity in a manner of speaking increases as radiant energy decreases.

24

The phenomenon of the consolidation of mass appears most dramatically in newly evolving stars. However, the same phenomenon applies also to smaller units of mass such as planets. Planets are not self-luminous bodies as stars are. Still, the same gravitational effects are at work seen in the compacting of the mass of a planet around its gravity center. For instance, the gravity at the surface of a compact mass 12,800 kilometers (7,900 miles) in diameter such as Earth is about the same as an amount of material almost 100 times as much spread over a less dense volume that is 120,000 kilometers (74,600 miles) in diameter such as the planet Saturn. Figure **2.7** illustrates the relative size relationship of these two planets — Earth and Saturn — each very different in their distribution of mass. Saturn has 95 times as much mass as Earth, yet an object weighing 100 pounds on Earth would weigh about the same amount on the surface of Saturn, in this case 107 pounds, because the

2.6. The Horse Head Nebula.

2.7. Saturn and Earth. Gravity at the earth's surface is roughly equal to that of the planet Saturn, although each of the two planets contains a very different distribution of mass. Photo of Saturn by Hubble Space Telescope.

mass of Saturn has not yet compacted about its center of gravity to the same extent as Earth. Weight is approximately the same because the surface gravity of the two planets is about the same although Earth's surface is nine and one-half times nearer its center of gravity than is the surface of Saturn whose mass is spread over a much larger volume.

25

Gravity is the unseen force that shapes the cosmos. Without it, there would be no universe or no universe, as we know it. Gravity, mass, and radiant energy appear to be interrelated, but no satisfactory theory yet exists to unify gravity with other energy behavior. Einstein searched in vain for a unified field that would explain all energy, and he never found it.

Remarkably, gravity always originates at a center of mass, a theoretical, indefinable point at which there is nothing. It is possible to measure gravity, feel its effect, see its results, and still not know how to break its laws or alter its structure without changing the mass that generated it. Gravity is the most enigmatic of all invisible energy.

The universe is full of trillions and trillions of objects, each with its own gravity center, and each one competing for available matter. Some of these objects coalesce, draw together, and combine their masses into larger bodies and systems. Planetary systems coalesce around stars, stars grow into clusters of stars, clusters into galaxies, galaxies into clusters of galaxies, until at some point the competing gravitational fields come into balance. Objects suspended in space — systems within systems — move through the universe locked onto each other in a mutually supportive, harmonious network of gravity centers. These centers are large scale and small; some are like our solar system or on a smaller scale like Moons around planets. The remarkable photograph in Figure

2.8 depicts the balance of gravity in one such familiar system, our Earth and its Moon. The barren ochre Moon seen from outer space floats suspended in orbit above an ocean-encircled Earth, their sunlit profiles standing in stark contrast to the blackness of outer space. If seen from Earth, the Moon, which is in the foreground in this photograph, would appear as a half-moon or in its three-quarter phase depending on the vantage point from Earth.

The Moon and Earth are a gravitational system, assured as enduring companions for millions of years to

2.8. Earth and moon photographed by the Galileo satellite.

come. Moreover, the solar system overall is a stable, harmonious system sustained by the Sun's mass and gravitational field, nurtured by its visible and invisible radiation, and linked across great distances by the delicate balance and motion of the planets that uniquely define it.

Seeing Earth alone in the blackness of space beside its single, lifeless Moon is spiritually uplifting for its unparalleled beauty, and yet sobering. Inspiring because mortals see Earth as it is and not as once imagined when only the Great Soul could see it as a creation beyond human scale. Earth is at once peaceful and yet fearful in its isolated uniqueness because Earth is dependent entirely on the Sun. For all of humanity there is nothing to replace it. Were it not for the reassuring illuminating warmth of the Sun, Earth, the Moon, and the planets would disappear into the everlasting night of the invisible universe.

Mauro Gandolfi. *Allegorical Figure of Astronomy*. ca. 1798.
Pen and brown ink, brush with blue, rose, yellow, and brown wash, over black chalk, 8 x 7 15/16 in. (20.3 x 20.2 cm.) Rogers Fund, 1961.
The Metropolitan Museum of Art.

IV

GRAVITY SYSTEMS AND ORBITS

26

As mass builds in the vacuum of space, it begins to rotate. As gravity draws the mass into a sphere, the sphere spins faster and faster. The rate of spin depends on the density of the object, or its mass-to-volume ratio. The greater an object's mass, the higher is its potential for rotation. A high mass to volume ratio, or high-density, will result in rapid rotation while a low ratio causes slower spin.

One example of this phenomenon is the planet Saturn. With its low mass-to-volume ratio, Saturn rotates less than one-half the speed of Earth. Because of its relatively high mass, however, Saturn has the potential eventually to speed up as its mass condenses closer to the planet's center of gravity. The principle of the conservation of energy enables a set amount of energy to spin the same mass at different velocities depending on the volume of space the mass occupies. A more compact mass will have a higher velocity of rotation than the same mass spread over a larger area.

This is like a spinning figure skater who at the beginning may spin slowly but goes faster and faster as arms and legs draw closer to the skater's body into a more compact, denser space. In this case, the body's volume masses toward the skater's center of gravity, causing the velocity of spin to speed up. Spin will continue at the faster speed until the arms again extend or until the friction of the ice and air drags the skater to a stop. In space, there is no friction. The spin and momentum of celestial objects change only by their internal transformation or by their occasional encounter with other celestial bodies, speeding up and slowing down as energy and material exchange and mass redistributes to create a different relationship of mass-to-volume ratio.

27

The near side of the moon

If a concentration of mass exhausts all or most of its radiant energy and is within the gravitational field of a substantially larger object, it may slow and eventually cease to rotate. For example, the Moon, which has no radiant energy, does not spin at all. That is why we cannot see the dark side of the Moon from Earth. While the Moon possibly did rotate at one time, Earth's gravity has locked onto it, pulling at its slightly more massive near side, which is always toward our view. Most Moons of other planets also show only one side to their planets because like our Moon those Moons, too, have masses that are only fractions of the masses of their hosts, and they, too, float suspended in arrested motion.

The Moon is relatively small. At 3,475 kilometers (2,160 miles) in diameter, it would fit comfortably between the cities of Los Angeles and Washington, D.C. The Moon has only 1.2 percent of the mass of Earth, with one-sixth of its gravity. Therefore, the Sun pulls the Moon's lesser mass off center, just as the Sun's gravitational pull has helped to distort the shape of Earth in the form of a bulging equator. Like an out-of-balance tire that eventually stops rotating with its heavy side down toward Earth, the Moon is an out-of-round sphere also pointing toward Earth.

To say that only one side of the Moon is visible is not exactly accurate. Because of its size and nearness to Earth, different angles of view from positions on either side of Earth make it possible to see more of the Moon's surface than would be possible from a single vantage point. In addition, since the Moon travels around Earth, very nearly the same experience is obtainable by observing the Moon as it rises, and by following it across the sky to the other horizon. Either way, about 10 percent more of the Moon's surface is visible than would be if it stayed directly overhead. Because the lopsided face of the Moon always points toward the center of Earth, a view from one side of Earth will show more lunar features than a view taken from a different angle. The exaggerated angles in the diagram of this phenomenon (**3.1**) show how the angle of vision changes from horizon to horizon. Observations taken at points ***A*** and ***B*** reveal about 60 percent of the Moon's surface.

3.1. Diagram of the light side of the moon.

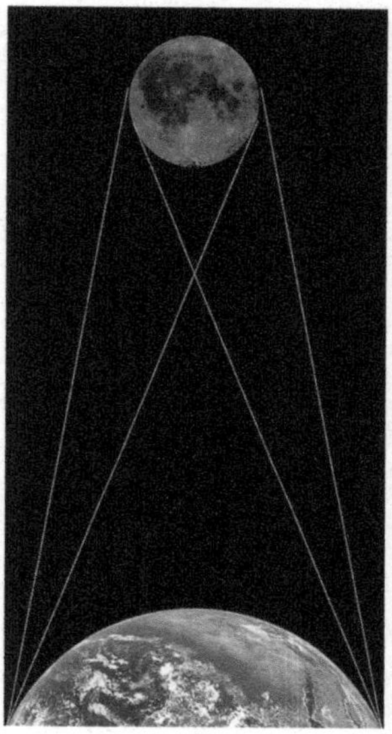

A ***B***

Earth itself may someday become sufficiently lopsided that it, too, will stop rotating, perhaps billions of years from now, locked in one position forever facing the Sun. This process may already be underway. The rotation of Earth is gradually slowing down.

Earth is 0.34 percent larger in diameter at its equator than it is at the poles. This is in part due to its rotation but is due also to the gravitational pull of the Sun. The Sun exerts greater force on the part of Earth that is closest to it — the equator — than it does on the poles. These two conditions: the pull of the Sun and the planet's rotation, combine to flatten Earth's spherical shape and cause it on average to be about 43 kilometers (27 miles) larger at the equator. At the same time, because of its oblate shape, Earth's gravitational force is slightly less at the equator than at the poles.

28

Orbits

The balance of a cosmic system is in the motions of the objects within the system. When the individual motion of an object, a planet for example, closes on itself, it is in orbit. Orbits are circular or elliptical paths around a body whose gravitational force is sufficient to balance the momentum of the orbiting objects. The most familiar of such systems is, of course, the solar system in which the planets orbit the Sun.

An orbit is a line tracing the path of an object traveling through space at a specific velocity. When the path of an object passes into the gravitational field of another object, gravity bends the trajectory of the line of the lesser object into a curve to form an orbit around the more massive object; like the Moon around Earth.

At some time in the distant past, Earth's gravity captured the Moon and curved its path into its present orbit. This is not to say, of course, that the Moon came from outside the solar system. One of the open arguments in modern astronomy is whether Earth captured the Moon as the two passed each other, or whether the Moon formed out of Earth itself in a collision with another object. Nevertheless, astronomers generally agree that the Moon came from within and not outside the solar system, whether as a collection of solar debris left over from the formation of the solar system or as a fragment of Earth

29

The success of an orbit depends on gravity and the velocity, or momentum, of the orbiting object. Gravity and momentum must be in precise balance. Momentum is always in a straight line and in a direction opposite the force that initiated the momentum.

The fact that an orbit represents a straight line of momentum manifests in the slingshot, an ancient device used in hunting and warfare. A stone placed in a holder at the end of a cord whirls rapidly in a circular motion. This circular motion is the stone's angular momentum. As angular momentum builds up, the stone travels faster and faster. When the stone releases from the slingshot, it does not continue its circular motion as one might expect. Instead, its instantaneous velocity immediately assumes a straight-line trajectory that is a tangent to the circular path. A tangent means the line of travel is perpendicular to the radius of the circle and touches the curve of the circle of angular momentum at only one point in the circle's path. Figure **3.2** illustrates the geometry of angular momentum and instantaneous velocity. The force of a slingshot projectile then comes from momentum built up by its circular path. The projectile releases toward its target precisely at the point at which the line to the target ***o*** is a tangent to the circle of the motion ***p***; that is, when the line is perpendicular to the circle's radius. Where the line is tangent to the circle determines an object's direction when released from its circular motion. In the example of a slingshot, the release point determines whether a stone strikes its target. Instantaneous velocity, whereby the angular momentum of the circular path transfers immediately to the straight path of the projectile, characterizes the straight line of the projectile.

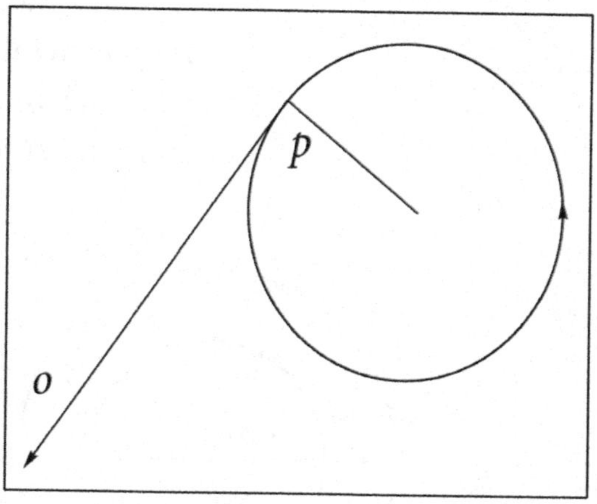

3.2. Instantaneous velocity. The principle of the conservation of energy transfers circular motion—called angular velocity—into instantaneous, or straight, velocity.

This same behavior of force is at work in orbiting bodies. Imagine if Earth's gravity suddenly switched off. The Moon and all the artificial satellites in orbit around Earth would each go zipping away on a straight line until they encountered new places of gravitational force. Each one would immediately resume a straight line of travel and, depending on its direction, could wind up in orbit around another planet, collide with a solar body, or perhaps exit the solar system altogether. Planets and satellites, unlike slingshot projectiles, do not aim at other objects. Rather, gravity transforms their straight-line velocities into graceful paths that define in perfect harmony the orbits of the spheres.

30

The Barycenter

An object in space moves through the universe with a primeval momentum given to all gravity centers. In the course of finding the balance within a cosmic system like the solar system, for instance, the momentums of objects bring them into proximity with one another, causing the individual rotating bodies to begin orbiting around a common center of balance. This point of gravitational

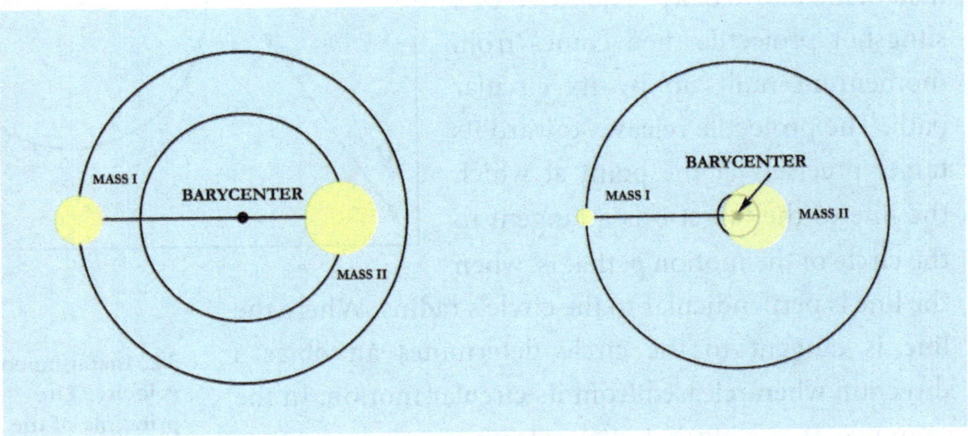

balance in a cosmic system is the Barycenter. Objects orbit this center of balance in the same period. This is not the same, as a planet orbiting the Sun. Everyone knows that the planets all have different periods that it takes for them to complete one trip around the Sun. Instead, the Barycenter refers to a singular relationship between two bodies, such as the Sun and one of its planets.

3.3. The Barycenter

The Barycenter appears most clearly in a binary star system where two stars of similar mass orbit each other (**3.3**). The line joining two gravity centers passes through a point of balance set at a distance relative to the masses of the two objects. Two objects of equal mass will appear to orbit each other because their Barycenter is located precisely between the two bodies. When the two objects are of very different masses, however, the Barycenter will be at the center of gravity of its two components and may be even inside the larger of the two. In the first instance (left diagram 3.3), the Barycenter is located between the two stars relative in distance to their respective masses. It will take each star the same amount of time to complete one orbit around the Barycenter. In many cases, however, a Barycenter will mark a more eccentric balance. A less massive object in orbit around a more massive one, such as the Moon around Earth or Earth around the Sun, will cause the barycenter to be closer to the more massive object in order to be in balance, instead of being equidistant between two objects of the same mass.

31

Alternate placements of the Barycenter relative to the masses of two orbiting objects show the eccentricity in their orbits. The Barycenter of any two objects of significantly different masses, one very large, one relatively small, actually falls within but may not be exactly at the center of the larger more massive host object (right diagram 3.3). For example, the Barycenter of the Earth-Sun balance is within the radius of the Sun, while that of the Moon-Earth paring is within Earth. This eccentric point of balance results in a detectable wobble in the host body. The wobble represents the larger body's tight rotation around the Barycenter. The amount of a wobble is dependent on the placement of the point of balance determined by the size of the smaller orbiting body. The larger the orbiting objects relative to the host, the more wobble in the host. The planet Jupiter, for example, is large enough to cause the Sun to exhibit a detectable wobble. This is also one way to discover a large but invisible planet orbiting a distant star. We may not see a planet outside the solar system, even with a telescope. Nevertheless, the wobble seen in its host star predicts the presence and size of a new planet.

When the mass of the orbiter is sufficiently large to cause the Barycenter to be outside the host, the two objects will appear to orbit around an invisible common center, causing them to appear to orbit each other.

The Barycenter is a modern discovery. If early astronomers had known about it, they would have known that the Sun could never orbit Earth as once believed, thus advancing astronomy and physics by a thousand years.

32

Elliptical orbits

Orbits are rarely circular in shape. They are, for the most part, usually elliptical, that is, longer than wide. The angular momentum of an elliptical orbit behaves essentially the same as in a circular orbit, except the major gravity point in an elliptical system, instead of being at the center of the system, as in a circular orbit, will be always at one of two focal points that define the ellipse. In Earth's elliptical orbit around the Sun, the Sun is at one focal point. Similarly, Earth is at one of the focal points of the Moon's elliptical orbit.

In an elliptical orbit the orbiting body, Earth for instance, is at varying distances from the primary focal point, in this case the Sun, depending on where Earth is in its 365-day orbit. The changing distance of Earth from the Sun affects the planet at different times of the year. The elliptical path of an object also contributes to the Barycenter effect between the object and its host. The slight shifts of the Barycenter within the Sun, for instance, adds to the dynamics of the Sun's thermal activity in a way related to Earth's tidal changes due to the gravitational force of the Moon in its elliptical orbit.

33

The German astronomer Johannes Kepler was the first to describe orbital behavior in relationship to the mathematical geometry of the ellipse. Among other things, he discovered that orbiting bodies sweep out equal areas of an ellipse in equal time (**3.4**), meaning that orbiting objects must speed up as they approach the primary focal point of the ellipse and slow down as they recede away from it. Unlike the velocity of an object in a circular orbit, where speed is constant, an object in an elliptical orbit must speed up and slow down according

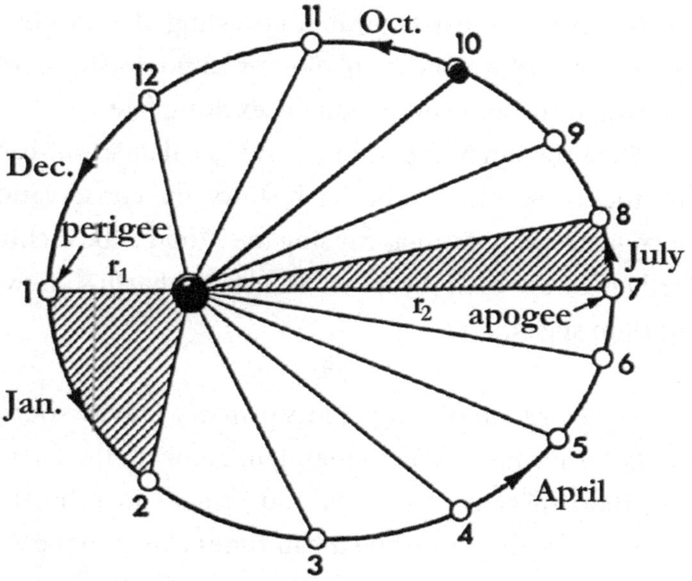

3.4. The elliptical orbit of earth. Earth is farther from the Sun in July than in December.

to its proximity to the controlling gravity host. In the 12 months that it takes Earth to complete one revolution around the Sun, the planet's velocity varies from its highest speed in early January when it is closest to the Sun, to its slowest speed in July when it is at its farthest point from the Sun. Earth is three million miles closer to the Sun in January than July. If Earth's orbit were circular instead of elliptical, the average temperatures in the hottest and coldest months would change about two degrees from current averages.

34

Satellites

A significant change in either the mass or velocity of an orbiting body will alter its orbit, or destroy the orbit all together. This is why low orbiting artificial satellites at altitudes below 160 kilometers (100 miles) eventually fall back to Earth. There is just enough molecular friction in space at low-orbit altitudes to gradually slow a satellite's speed until eventually it comes crashing down, burning through Earth's atmosphere.

The problem of maintaining a steady orbit is not limited to artificial satellites. If, for instance, the Moon should increase in speed beyond its current velocity, it would have to adjust its orbit or go sailing off on its own, possibly to orbit another planet, or to go crashing into the Sun, or to some other equally exciting fate.

The problem of the Moon's orbital balance is already in process. As the spin of Earth slows, the conservation of angular momentum is causing the Moon to be inching farther away each year to adjust to Earth's slower rotation speed.

35

The closer an object is to its primary gravity source, the faster it must travel to maintain an orbit; the farther out, the slower it can travel and remain in orbit. For example, Earth is nine and a half times closer to the Sun than the planet Saturn, so Earth must orbit the Sun at a speed three times that of the ringed planet. This velocity-to-distance relationship is the reason why placing manmade satellites in orbit is such tricky business. A satellite must reach a precise altitude and velocity relative to its mass, or weight, to achieve an orbit (**3.5**). A satellite launched from Earth must first achieve a

3.5. Orbit of an artificial satellite.

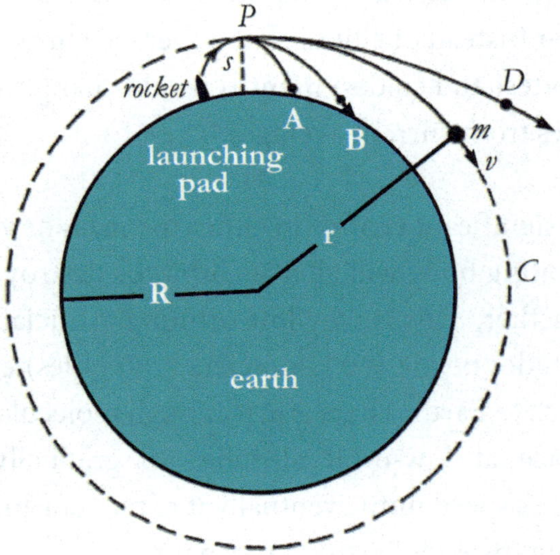

velocity of 40,300 kilometers (25,000 miles) per hour just to escape the gravitational pull of Earth. Anything less and it will come tumbling back down. The description of a satellite's orbit around Earth compares to that of an object perpetually falling toward Earth but never reaching it. The orbital velocity v of a satellite is dependent on the thrust P of the launch vehicle. Too little thrust and the projectile will fall back to Earth at A or B; too much thrust and the satellite will escape Earth completely. Physicists rely on Newton's Laws of Gravitation and Motion to calculate the exact amount of velocity required, thus achieving orbit C. The force of gravitation exerted by Earth on a satellite of mass m is inversely proportional to the square of the radius of Earth. A satellite will always fall back to Earth unless it attains sufficient velocity to carry it the right distance beyond Earth's atmosphere and gravitational field. At that point, as the satellite falls back toward Earth, Earth's curvature drops the planet's surface from underneath the falling satellite so that it is constantly falling toward Earth without ever hitting it.

Once beyond the atmosphere, a satellite can travel at about 29,000 kilometers (18,000 miles) per hour and remain in orbit. Because it is so close to Earth, an average required orbital velocity of a satellite is nearly ten times that needed by the Moon whose orbit is at a distance from Earth seven hundred times that of most artificial satellites. The Moon is traveling at an average velocity of 3,700 kilometers (2,300 miles) per hour.

36

At the beginning of space flight, satellites were very small. Rockets that could reach the necessary escape velocity of Earth's gravity and achieve orbital speed were capable of boosting only tiny payloads into space. The USSR Sputnik was the first artificial satellite successfully sent into orbit. Sputnik is Russian for the word "satellite."

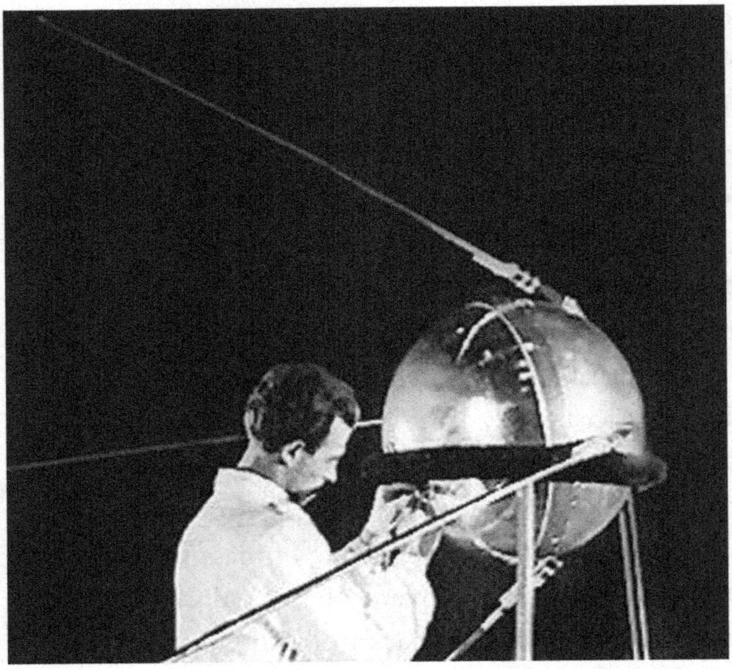

3.6. Sputnik. The first artificial satellite.

Launched in 1957, Sputnik weighed 84 kilograms (185 pounds) and marked a milestone in space exploration. Figure **3.6** depicts its scale next to a Russian technician.

From its beginning in the 1940s, rocket science developed rapidly. The first U.S. space station, Skylab — launched just 16 years after Sputnik — weighed 82.5 tons (**3.7**). Placed in orbit in 1973, Skylab stayed up for more than six years before falling back to Earth. In the comparatively brief span of a few decades, it is possible today routinely to launch objects of several tons. The Space Shuttle, for instance, weighed 75 tons.

Compared to the magnificent scale of the planets and their Moons in the solar system, artificial satellites seem almost insignificant. Yet the discovery and application of the laws of physics that govern orbiting bodies are among the great intellectual and practical achievements in the history of astronomy and astrophysics. We see satellites as points of light among the stars. Each one moving silently across the celestial plane, first one, and then another to suddenly and mysteriously disappear as its reflective covering passes into the shadow of Earth to reappear on the horizon again only minutes later.

3.7. Skylab. The first US space station was in orbit from 1973 to 1979. NASA.

Hurricane Josephine seen by the space shuttle orbiter Challenger. 1984. NASA.

V

THE GOSSAMER EARTH

The atmosphere is composed primarily of nitrogen and oxygen with traces of helium and other gasses. Based on temperature, the atmosphere divides into five layers, viz., troposphere, stratosphere, mesosphere, thermosphere, and exosphere (**4.1**). Most of the atmosphere is concentrated in the troposphere, roughly the first 10 kilometers (6 miles) above the surface of Earth. Our experiences of weather occur in this layer.

4.1. Layers of Earth's atmosphere. The yellow line shows air temperatures at increasing heights. ***Encyclopedia Britannica.***

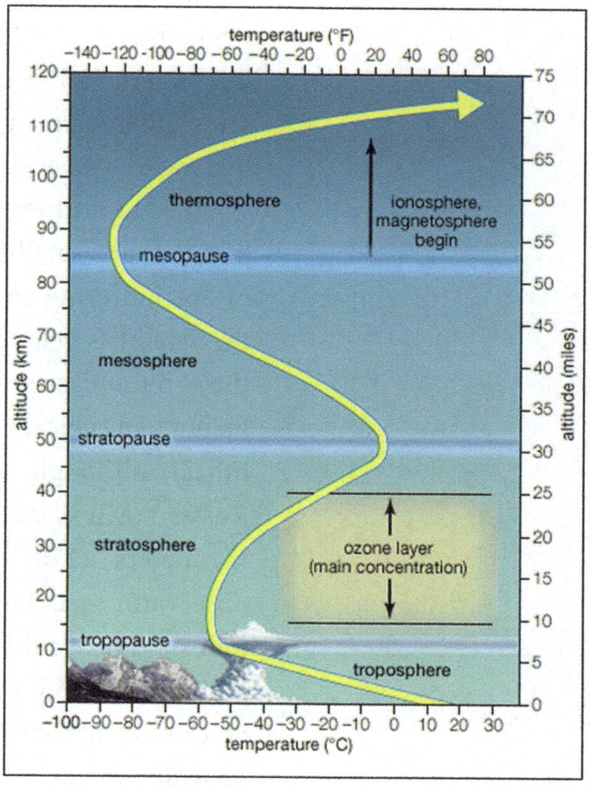

The stratosphere lies above the troposphere and extends to an altitude of around 50 kilometers (30 miles). The stratosphere's ozone layer intercepts the Sun's invisible ultraviolet light. Only about 50 percent of the Sun's energy is able to penetrate this layer to reach the ground; the rest falls away reflected or absorbed. Above the stratosphere is the mesosphere, up to an altitude of about 80 kilometers (50 miles).

Altogether, the first 80-100 kilometers (50-65 miles) of the atmosphere we call the ecosphere because this is where most of the life-sustaining gases exist and where all climatic changes occur.

Temperature in the atmosphere goes up and down as a function of altitude. Temperature decreases with altitude in the lower troposphere, then increases in the stratosphere, before rapidly decreasing again with altitude in the mesosphere. Going higher still, temperatures rise a second time in the overlying layer named the thermosphere, out to about 600 kilometers (375 miles), before entering the exosphere, the last identified layer of the atmosphere where temperature loses its customary meaning as gasses thin out and give way to the absolute coldness of outer space.

The envelope of the atmosphere is a blending of its different layers. Elements of each layer exist in contiguous layers. Oxygen exists detected as far out as 1,100 kilometers (685 miles); helium is present out to 3,500 kilometers (2,175 miles).

At the same time, atmospheric pressure decreases, as the air grows thinner farther out into space. Already negligible at 150 kilometers (93 miles), beyond 3,500 kilometers (2,175 miles), all traces of the gases of the atmosphere are gone, replaced by the interstellar hydrogen of space.

It is at altitudes of 160 kilometers (100 miles) and higher that a satellite must orbit to escape the possibility of encountering enough atmospheric friction to reduce the satellite's velocity. A loss of velocity causes decay of the orbit and probable loss of the satellite. The orbiting height of a typical satellite is around 375 kilometers (235 miles), although some fly lower, some higher. The Russian Mir space station, for instance, flew at 272 kilometers (170 miles), while the Chandra X-ray observatory satellite orbited at 34,555 kilometers (21,575 miles). The Hubble Space Telescope orbits at 560 kilometers (355 miles), and the International Space Station at 400 kilometers (250 miles) above Earth. To provide an Earth perspective of these distances, a commercial passenger jetliner cruises in the troposphere at about 9 kilometers (5.5 miles).

Synthetic satellites orbit in space only a few hundred kilometers above Earth, reminders that it is a fragile Earth we inhabit. Satellites are visible with the unaided eye racing across a clear evening sky just beyond the atmosphere and marking the boundary between outer space and the thin blanket of gases that sustains life. Seen from space the atmosphere is a narrow blue band glowing on the distant horizon against the empty blackness of space (**4.2**). The International Space Station

floats above the blue haze of the atmosphere seen in the distance in a photograph taken from a Russian Soyuz spacecraft. This was the first picture taken of a shuttle docked to the International Space Station.

4.2. International Space Station.

Life on Earth has a precariously thin biosphere. The biosphere is that part of Earth and its atmosphere where living things exist. The breathable atmosphere extends out less than eight kilometers (five miles) before human lungs need assistance. The sustainable atmosphere is barely 0.16 of Earth's diameter. Imagine a 13-story building covered by a one-inch roof and that is the relationship of the viable atmosphere to Earth, like a ball five feet in diameter wrapped in a single layer of tissue paper. This thin, invisible veil of gasses sustains life and protects it.

38

In the vicinity of earth

Very little new material remains near the solar system to add to its mass. It is about as massive as it will ever be. However, mass rearranges occasionally in collisions of sometimes spectacular and cataclysmic dimensions. Many such collisions have happened in the past, and many more are likely to occur in the future, some with profound consequences.

Over a century and a half after their discovery, the fate of the dinosaurs remained unknown. Once the prime inhabitants of Earth, they suddenly disappeared from existence, leaving fossil remains that served only to confirm the abruptness of their departure. Then in 1980, scientists announced that the extinction of the dinosaurs resulted from global climatic changes brought on by the

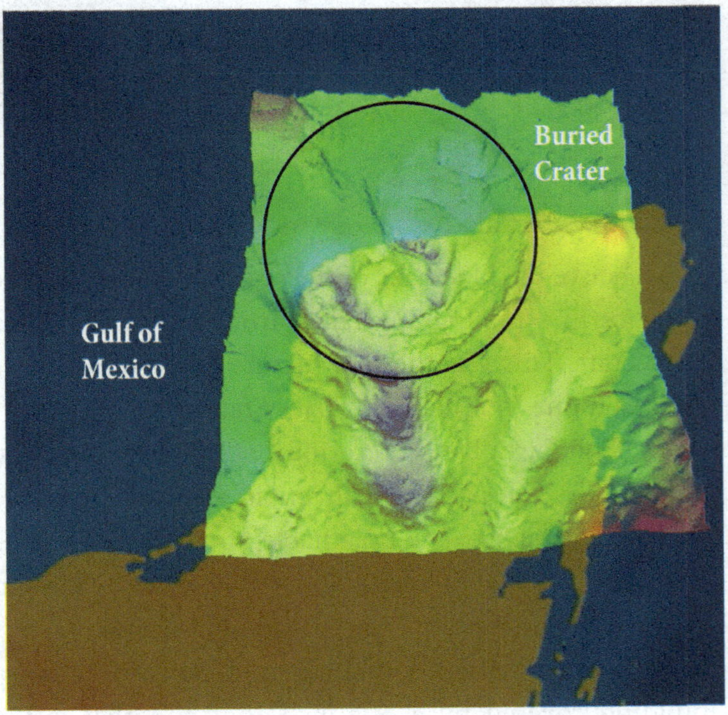

4.3. Chicxulub Crater, Yucatan Peninsula, Mexico.

impact of a large extraterrestrial object that collided with Earth. In 1999, another announcement hailed the discovery of the site of the object's impact on what is now the Yucatan Peninsula, on the east coast of southern Mexico. A magnetic field image revealed the massive Chicxulub Crater of the extinction collision (**4.3**). The image, created from gravity and magnetic field variations, covers a large portion of the Yucatan Peninsula's tip. The actual crater lies buried from view beneath hundreds of meters of sediment. The crater, 180 kilometers (112 miles) across, marks the location of the collision, thought to be with an asteroid measuring 10 to 20 kilometers (six to 12 miles) in diameter. The force of its impact ejected billions of tons of material thrown into the atmosphere. The whole planet plunged into darkness from which half the species then on Earth never emerged.

Known as the Cretaceous-Tertiary extinction, the cataclysm was greatest in North America. The asteroid allegedly approached from the southeast and hit at an angle, throwing debris toward the northwest that created a parabola-shaped kill zone over western and Central America. The kill zone extended as far to the east as the Appalachian Mountains and west beyond the Pacific shore, all the way to Siberia.

The dinosaurs were not the first terrestrial inhabitants wiped out by a collision with an asteroid or comet. There have been at least ten such extinctions since Earth's beginning, five of them particularly devastating ones. Two hundred fifty million years ago, for example, an event similar to the one that killed the dinosaurs led to the extinction of 90 percent of all marine creatures living at that time and 70 percent of those living on land. The object that struck Earth then would have been very large, too, at least as large as the one that struck the Yucatan 195 million years later. Capable of releasing energy equivalent to a million times the worst recorded earthquake, it triggered volcanoes whose ash and dust plunged the world into centuries of dark and cold. No one yet knows where this earlier collision occurred.

Such a cataclysmic collision satisfied the scientific community as the long sought solution to the extinction of the dinosaurs, one of the great mysteries of the past. Not much thought at the time went into the relevance of such events to the present. The implications that such a catastrophe could occur again received little notice, until 1994. That was when the world witnessed the broken comet Shoemaker-Levy 9 slam directly into the planet Jupiter, leaving dark Earth-size holes trailing across the planet's atmosphere. In a remarkable image (**4.4**), eight impact sites were visible. (Some impact sites were close together and appeared as a single black spot.) The

4.4. Jupiter and Shoemaker-Levy 9. Dark spots mark the impacts with the disintegrating comet Shoemaker-Levy 9. Hubble Space Telescope.

impacts strung out across the planet's surface because of Jupiter's rotation. Parts of the broken comet struck in a string of impacts at different times. The largest fragment hit Jupiter with a force of six million megatons, enough force potentially to destroy Earth if a similar collision were to occur.

The collision of Jupiter and Shoemaker-Levy 9 was the first encounter of two large solar system bodies ever observed. Almost immediately, initiatives stepped up to search for and identify objects in space that might pose future hazards to Earth. There was a resurgence of interest in tracking such objects for their potential threat to Earth. A spate of poorly conceived and badly documented science fiction films inundated the public, attempting to capitalize on the immediate interest in extraterrestrial collisions. Like most sensationalism, however, public response to doomsday scenarios soon quieted, awaiting redirection to the next phenomenon. Fortunately, the search for extraterrestrial hazards did not stop. It continues as an important part of space investigation.

39

Bolides

Many kinds of objects in and outside the solar system are capable of colliding with the planets. These belong to a class of objects called bolides: meteoroids, asteroids, and comets. A bolide is a fireball in the atmosphere. However, since all objects entering the atmosphere become fireballs, bolides and the collection of objects represented by the term are the same.

40

Meteoroids are celestial objects ranging in size from dust particles to objects the size of small asteroids and may come from a variety of places. They may come from within or outside the solar system and are the most common kind of collision-bound matter. Meteoroids appear as meteors, or shooting stars, when entering Earth's atmosphere (**4.5**). Most of them burn up in the atmosphere; however, some do not and they strike Earth.

Many thousands of meteors enter the Earth's atmosphere every year. Most of them disintegrate into dust-like particles; scientists estimate that Earth receives a billion such particles every day. Large meteors are visible entering the atmosphere at an altitude of about 89 kilometers (55 miles) and usually break up in the first 16 kilometers (10 miles). A meteor bolide can be bright enough to illuminate the landscape, and the roar of a large descending meteor may audibly resonate over many miles. People who have been near large meteor impacts describe the thunderous roar as truly frightening.

4.5. Bright bolide meteor over Black Mesa State Park. Photo by Howard Edin.

Meteors are of various origins including debris from comets and other kinds of bolides. Some meteors, for example, mark the point at which Earth passes through the orbit of an asteroid or comet. Debris left by comets accounts for most of the meteor showers, those unique occasions during the year when for several hours hundreds of such events bombard the sky. Such encounters occur throughout the year but are most frequent in the month of November. In its orbit around the Sun, Earth passes annually through one or more comet debris paths.

As Earth passes through a trail of debris, meteor showers show where the comet crossed the orbit of Earth. While very beautiful in their light displays, meteors from comet debris, when one thinks about it, are a sobering reminder that a comet was at one time within the collision path of Earth.

When a meteoroid survives the trip through the atmosphere to reach Earth's surface, the stony or metallic material remaining is a meteorite. A meteoroid is a meteoroid in space, a meteor in the atmosphere, and a meteorite if it reaches Earth. Made up of different materials, meteorites with glassy characteristics are *tektites*, stony ones are *chondrites*, and those made mostly of metal are simply *irons*. Tektites come from nearby, blasted out of the Moon's surface, for instance, while stony and iron meteorites usually come from outer space. About 95 percent of all meteorites are of the stony type.

Museums preserve thousands of meteorites retrieved from Earth's surface. As many as 2,000 meteors a year survive without disintegrating to make it to Earth's surface as meteorites. Most of them are smaller than ten kilograms (22 pounds), although one that survived the trip through the atmosphere weighed 54,000 kilograms (60 tons) and still rests where it fell in Namibia on the southwestern coast of Africa. Another that weighed 31,000 kilograms (34 tons) landed in Greenland. These two objects, however, are exceptions. Most surviving meteorites weigh only a few kilograms at most.

41

Asteroids are a second type of bolide comprising large, irregular shaped bodies with diameters up to several hundred kilometers orbiting the Sun. The first asteroid discovered was Ceres, found on January 1, 1801. At 914 kilometers (568 miles) in diameter, Ceres is the largest asteroid discovered. Ceres is roughly one-fourth the size of the Moon but not reflective enough to be visible to the naked eye. Only one asteroid, Vesta, is bright enough to see with the unaided eye, although several asteroids are visible with binoculars or a small telescope. Vesta is the third largest asteroid, with a diameter of 525 kilometers (326 miles).

In 1993, the spacecraft Galileo came within 2,400 kilometers (1,500 miles) of an asteroid named 243 Ida

(**4.6**) to take a close-up photograph of its heavily cratered, irregularly shaped body. Ida is comparatively small at about 56 kilometers (35 miles) in length. A remarkable discovery by the Galileo expedition was that Ida has a Moon, a 1.5-kilometer (about one mile) natural satellite named Dactyl. It is the first natural satellite of an asteroid detected, seen in this photograph of Ida just to the right of the asteroid. Dactyl most likely came about blasted away from Ida as the result of an impact with another asteroid or comet.

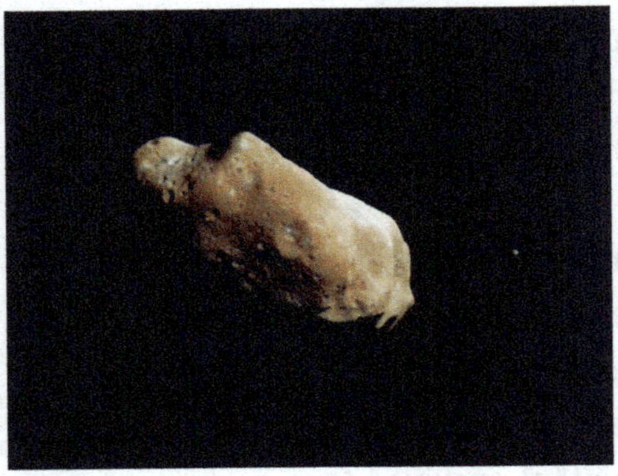

4.6. Ida and its moon, Dactyl. NASA.

Asteroids are large enough that a satellite actually landed on one. In 2001, the satellite space probe known as NEAR Shoemaker landed on the asteroid Eros after a yearlong orbit that took the satellite 315 million kilometers (196 million miles) from Earth. Eros is a long rock 34 kilometers (21 miles) in length (**4.7**), approximately the length of Lake Tahoe, Nevada, and four times the estimated size of the Chicxulub dinosaur extinction object. NEAR descended softly to the surface of Eros and continued to broadcast radio signals after landing.

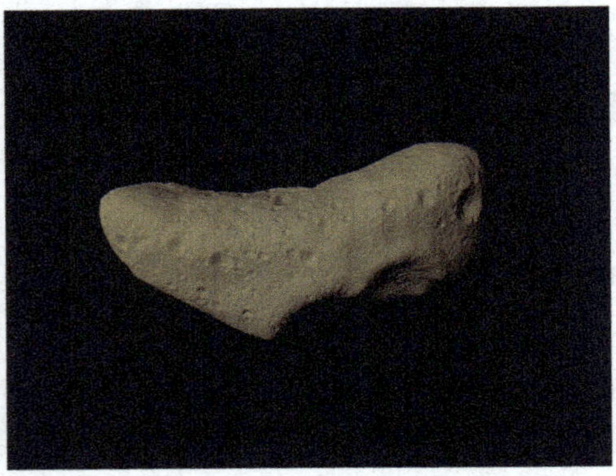

4.7. Eros. NEAR-Shoemaker. NASA.

Ceres, Vesta, Ida, and Eros pose no immediate threats to Earth. However, other asteroids do, like the asteroid Hermes, for example, which in 1937 came within 780,000 kilometers (485,000 miles) of Earth. Aside from the Moon, Hermes is the closest approach on record by an object of significant scale. A disquieting fact is that Hermes has since disappeared. No one knows where it is. Keeping track of asteroids is not easy. In 1998, yet another potentially dangerous asteroid became a concern, but it, too, vanished before observers could compute its orbit. Jupiter, whose strong gravitational field perturbs the asteroid belt with the potential to change asteroid orbits from time to time, further complicates the task of mapping the orbit of an asteroid.

Asteroids are a part of the solar system orbiting chiefly between the planets of Jupiter and Mars. A diagram of some of the known orbits of asteroids is in **4.8**. One group of asteroids, known as the Apollo asteroids, intersects the orbits of the inner planets. One of these, the asteroid Icarus, cuts across Earth's orbit to circle the Sun closer than does any other object of the solar system, closer than the planet Mercury. Icarus and

4.8. Planets, asteroids, and comets.

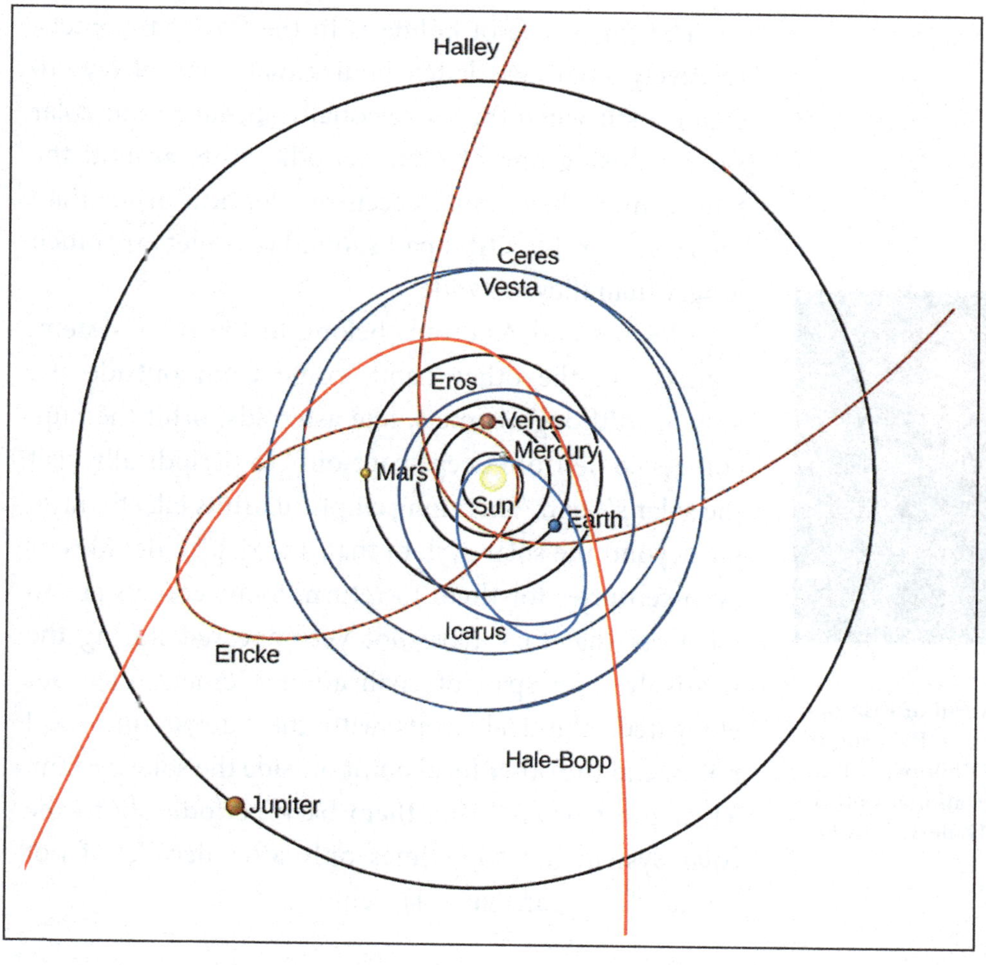

asteroids like it, with similar eccentric orbits, pose the greatest danger for a collision with Earth.

Of the different kinds of bolides capable of major collision damage, asteroids are the closest threat to hit Earth. Several thousand asteroids comprise the catalogue so far in our solar system; none appears at this time to be on a collision course with Earth. Nevertheless, while most asteroids pose no immediate threat of collision, a handful does. All together about 2,000 of these special objects are in the watch catalogue, some of them large enough to cause serious damage if they collided with Earth. About 200 of them have collision potential. Their orbits are erratic and appear to cut directly across Earth's orbit.

42

The third type of bolide is in the form of Comets; relatively small, gas-laden bodies that seem always to create a stir when they occasionally appear in our solar system during one of their periodic visits around the Sun. Comets have more eccentric elliptical orbits than those of asteroids. Orbit paths found in comets are much longer than they are wide,

4.9. Portrait of Edmond Halley. Painting by Richard Phillips, c. 1722. National Portrait Gallery, London.

Meteors and Asteroids belong to the solar system; comets, on the other hand, come from outside the system. Although comets, like asteroids, orbit the Sun, comets come from deep space only to periodically visit the solar system. Their long elliptical orbits take them so far beyond the solar system that it may take decades or even centuries for them to return. Some comets are so far away that they have not yet appeared during the relatively brief span of civilization. Comets trace out elongated, elliptical orbits with the Sun at one focal point, and the other focal point outside the solar system. Their great orbits bring them back periodically to the solar system but sometimes only after decades if not hundreds of years have elapsed.

English astronomer and mathematician, Edmund Halley (**4.9**), calculated the orbit of one comet that appeared in 1682 to coincide with its return through the solar system about every 76 years. It has varied over the last 2,000 years from 74.4 years to 79.2. Comet Halley travels in its orbit beyond the Sun 55,000 times farther out than Earth. Based on historical sightings of comets in 1456, 1531, 1607, and 1682, Halley concluded in 1705 that the sightings were of the same comet, which he predicted would appear again in 1758. When it did, Halley's posthumous public acclaim was such that the comet would forever bear his name. Comet Halley, last seen in 1986, will appear again in the year 2061. It has

4.10. ***Halley's Comet,*** **detail of the *Bayeux Tapestry.* Unknown artists, Saxon England or Norman France. c. 1073-88. Embroidered wool on linen, 51 cm x 70.5 m (20"x 231'). Musee de la Tapisserie, Bayeux, France.**

returned on schedule at least since 240 B.C.E., the oldest historical record of its appearance when the Chinese wrote about it. One of the earliest known pictorial representations of Comet Halley is in the *Bayeux Tapestry* (**4.10**). The *Bayeux Tapestry* chronicles the conquest of England by William of Normandy. Comet Halley was there in 1066 when the battle took place. Advisors considered it an omen to William's success.

The rare and unusual shapes of comets streaking through the sky made them of special interest to artists and other early observers of the heavens. The Italian artist Giotto depicted Comet Halley as the Star of Bethlehem in his painting, *Adoration of the Magi* (**4.11**) seen above the roof of the building. Meanwhile, recent history will remember Halley best perhaps in modern

4.11. Giotto (1267-1337). *Adoration of the Magi.* **1304-06. Fresco, 200 x 185 cm. (78 x 72 in.). Capella Scrovegni (Arena Chapel), Padua, Italy.**

times as the comet that was here when Mark Twain was born in 1835 and again coincidentally at his death in 1910.

4.12. Hale-Bopp Comet. Photo by Philipp Salzgeber.

Whether Halley's Comet was the celebrated Star of Bethlehem, as the artist Giotto pictured it is debatable. Comet Halley did visit in 10 B.C.E. and could have figured in the lore of the birth of the Christ child. Scholars place the birth of Christ at around 6 B.C.E.; close enough in time to be the celebrated guiding star of the annunciation. Still, no one knows for sure. It is certain, however, that the 10 B.C.E. appearance of Halley solidified the self-proclamation of Octavian as Caesar Augustus, divine emperor of Rome. The comet, some said, was the final departure of the soul of Julius Caesar, signaling that he, Octavian, was to rule henceforth empowered by the divine.

A more recent example of a comet to visit the solar system was the great comet Hale-Bopp. Hale-Bopp, discovered in 1995, appears in **4.12** at the peak of its display in 1997. Some 10 to 20 times more visible than most comets, Hale-Bopp is a periodic comet like Halley and as such visited our solar system previously some four thousand years ago. Hale-Bopp compares to the great comet of 1811, which history remembers in part because of significant events that coincided with its appearance, including Napoleon's attack on Moscow. No such historic event occurred during the more recent visit of Hale-Bopp.

A comet is composed of a solid nucleus of material surrounded by gas and water or ice. It may be as much as 75 percent ice. During the 1986 appearance of Comet Halley, for instance, observers discovered that Halley has a nucleus in the form of an irregularly shaped chunk of ice.

The atmosphere, or *coma*, of a comet characterizes its interaction with the Sun. Comet Halley, despite its low reflective core, has a bright head, or coma, and a tail produced by gas in its outer layer activated by exposure to the Sun. This is why most comets grow brighter as they approach the inner part of the solar system because they intercept the stronger radiation of the Sun. Upon close approaches to the Sun, a comet's outer layer may grow to be millions of kilometers in diameter. An elongated vapor trail accompanies this outer layer pointing away from the solar wind, perhaps reaching 150 million kilometers (93 million miles) in length. A comet's tail always points away from the Sun no matter what direction the comet is traveling.

43

Planetary odds

The potential for a major impact with an object from outer space varies as a function of the size of the object, and the odds of such an occurrence change accordingly. That is, we are far less likely to collide with something large enough to have global consequences than we are with small things that cause no damage because there are far fewer large things than small things to hit us. The interval between meaningful impacts capable of altering the planet in some way is about 10 million years. The prospect of being hit by an object of any appreciable size is unlikely, at least for the moment. The closest threat discovered so far is a 26-kilometer (16-mile) diameter asteroid whose orbit will intersect the orbit of Earth in the year 2880. That far out in time, scientists give it only a one in 300 chance of actually striking the planet.

Nevertheless, the future behavior of some objects (like the comet Shoemaker-Levy 9 that struck Jupiter) is impossible to assess. Comets lie at such great distances beyond the solar system that many times we do not know they are coming until we see them.

On the other hand, asteroids are part of our neighborhood, and we should be able to anticipate their danger. However, since 1994 several asteroids have been identified whose eccentric orbits and odd compositions have led astronomers to question whether asteroids are any safer than comets. Most astronomers agree that Earth will experience another cataclysmic event, hit again by something of significant size; it is only a matter of time.

Earth is a game of chance where we wait until a sequence of destructive events that mortals cannot stop are unleashed, when enough energy is released upon the planet to once more visit extinction on all but the most perseverant of life forms. Powerless against the darkness that shrouds these objects of fate; we do as humankind has always done, trust that there is a divine hand that guides terrestrial destiny.

Meanwhile, Earth follows its regular orbital path, bombarded constantly by small interplanetary particles that burn up in the atmosphere. We may see them as meteors — shooting stars, or falling stars. Any object of less than about 30 meters in size normally will disintegrate before it reaches the planet's surface.

Large objects are a different matter. We can expect some sort of impact from an object in the 30-meter (100-feet) size about every 100 to 300 years. One of the last such events from this class of objects occurred in Siberia in 1908 when a mysterious blast leveled several hundred square kilometers of forestland.

VI

THE MYSTERY OF TUNGUSKA

44

In the early dawn hours of June 30, 1908, an object roared into Earth's atmosphere with such force that it caused the planet to shake violently. People fell to the ground 60 kilometers away. Atmospheric changes occurred around the globe. Observers saw the fireball 600 kilometers (375 miles) away and heard it for 1,000 kilometers (625 miles).

The object came down in Podkamennaya Tunguska, a swampy, forested river basin in central Siberia and one of the world's richest coal reserves. A sparsely inhabited area, few people were in the region to witness the collision. Those who did see it described a large, extremely intense bluish-white fireball that streaked through the sky. It approached from the east and traveled across the morning sky in a westerly direction. As it plummeted toward the ground, it made a horrifying noise, ending in a tremendous blast. The record of seismic data from the explosion suggested that it was roughly equivalent to an atomic bomb of 10 to 20 megatons.

The aftermath of the encounter left an eerie, effervescent glow in the sky. Highly reflective dust caught the Sun's rays at altitudes high enough to illuminate a still dark Europe, making it light enough before dawn in London to read a newspaper. Nothing as strange had ever occurred in modern times before, yet it would be nearly 20 years before scientists visited the site in Siberia to see just how massive the devastation was.
A team of Russian physicists visited Tunguska in 1927. What they found were the charred remains of trees

5.1. The Tunguska site photographed in 1927 by Russian scientists nineteen years after a mysterious explosion left hundreds of square miles of charred trees, their tops pointing away from the center of the blast. *National Geographic.*

strewn outward for 30 kilometers in all directions (**5.1**), their tops pointing uniformly away from the center of the blast and defining an area of devastation 2,150 square kilometers (1,550 square miles) in size, half the size of the state of Rhode Island. At the center, where there should have been a massive impact crater, there was nothing, no crater and no evidence of an impact other than the flattened trees.

Based on the pattern of trees, and eyewitness accounts, scientists pieced together several scenarios to explain what may have happened that terrifying morning. The object, whatever it was, apparently came in at an angle of 30-40 degrees, a relatively small angle for such an object entering the atmosphere. The smaller angle of entry meant the object had to travel through more of Earth's atmosphere than if it had come from directly overhead at a steeper angle. This angle affected the explosion because nothing ever hit the ground. For there not to be an impact crater, the object, whatever it was, had to explode in midair. The thing was very large, at least in the 60-meter class and possibly the 200-300 meter category. Analysis of the charred remains of trees and the pattern of devastation indicated that the explosion occurred seven to eight kilometers (four to five miles) above Earth's surface.

The cause of the blast remains unknown. The best guess so far is for either a comet or asteroid. One set of experts argued that the collision involved the remains of comet Encke whose trail of debris Earth was passing through at the time. Encke comes into position with Earth about every ten years. It was here in 1908, and it links to two other explosions that are more recent than Tunguska, one over New Zealand and the other over Turkmenistan. Encke in its long history has broken into large chunks, and scientists think these pieces are large enough to cause massive atmospheric explosions. Other experts, meanwhile, believe the object that hit Tunguska was a stony meteorite or small asteroid. Tiny metallic spheres taken from the rings of surviving trees in the area of the blast are likely of extraterrestrial origin and are consistent with metals found in other meteorites. Still, there is no crater. For such a large object to disintegrate completely, leaving only sub-millimeter sized particles and nothing else, appears to defy logic and deliver in the process another of the mysteries of the universe to the planet's doorstep.

Numerous theories surfaced to explain this event. Some are fantastic. For example, the Serbian-American inventor Nikola Tesla, who is responsible among other things for the method of transmission of electrical power used today, as well as the electric motor, apparently believed that one of his experiments gone awry caused the destruction in Siberia. On the exact date of the Tunguska event Tesla allegedly tested a high-energy death ray aimed over the North Pole. He was convinced the device had overshot its mark. Many of the inventions of Tesla were ahead of their time, and this one certainly would fit what we know of Tesla. Nevertheless, astronomers have stuck to more mainline explanations about what happened. One theory, which sports a computer-generated model, appears to prove that even a very large object entering the atmosphere at a certain angle and speed could explode with sufficient force to fragment into nothing more than dust.

If the Tunguska event was an encounter with an extraterrestrial object of significant size, then we have entered the statistical era of a possibility, if not a probability, that another similar encounter is due.

The expected event already may have occurred. In 1996, some kind of large object exploded over northern Australia. Like the Siberian explosion, this one, too, happened over a thinly populated region. Witnesses were so convinced this time in their descriptions of what they saw and heard that the government launched an investigation to ascertain if the blast could have been of atomic origin, possibly by terrorists engaged in nuclear testing. Government experts ruled this out, but admitted that they could find no cause for such a massive explosion. Scientists, meanwhile, think some sort of object from outer space caused this event as well. Nevertheless, as in the case of the Siberian event, there was no impact crater found.

Strange atmospheric explosions have occurred at other times and places. In 1978 something unknown but powerful exploded over the Pacific Ocean. Meanwhile, Evidence indicates also that in the not so distant past the Incas of South America might have witnessed a 200-meter impact when an object struck Earth at a shallow angle, then skipped across the Peruvian landscape. So far, no references to the supposed event exist uncovered among Inca artifacts. Nevertheless, the geologic record appears to confirm it.

45

Craters

Most objects larger than 30 meters in diameter that hit Earth leave an impact crater. One such crater is located in the state of Arizona. The Barringer meteorite crater, discovered in 1891 near Winston, Arizona, is almost a mile across and 180 meters (600 feet) deep. Pictured in **5.2**, explorers identified it as an impact crater in 1920 when scientists guessed that an asteroid of about 45 meters (150 feet) created the crater. Geologists

5.2. Barringer Crater in Arizona. Most extraterrestrial bodies pose no immediate threat to the planets, although objects have collided with earth and other members of the solar system in the past. Lunar and Planetary Institute.

estimate the age of Barringer between 25,000 and 50,000 years. It stands as a concrete reminder that collisions with large extraterrestrial objects can and do occur.

Most craters of significant size are very old. The Barringer crater is an exception and is relatively young as impact craters go. Several older craters ranging in age from about 300 thousand to more than 300 million years exist.

One of the largest confirmed craters in the world) is a water-filled depression 100 kilometers (62 miles) in diameter in northern Quebec, Canada, pictured in **5.3** in a view from the International Space Station overlooking the St. Lawrence River. Named Manicouagan Crater, its size places it well into the class of catastrophic impacts.

This one occurred about 210 million years ago, and since then has been almost completely eroded by glaciation. The original rim is gone, leaving a circular impression to create Lake Manicouagan.

5.3. Manicouagan Crater. The International Space Station photographed Manicouagan Crater (right center) and the St. Lawrence River in Quebec while orbiting 258 miles above Canada. NASA.

Geologists are able to identify more than 120 terrestrial impact craters throughout the world, with an additional 80 or so other formations suspected also to be craters.

The discovery of impact craters happens all the time. One of the most recent ones discovered lies beneath the southern part of Chesapeake Bay on the Atlantic side of North America. An object three to five kilometers (two to three miles) in diameter slammed into the continental shelf, altering permanently the geological configuration of the region. At that time, 35 million years ago, the shoreline was near Richmond, Virginia, and the Appalachians were a tropical rain forest. The impact, whose crater lies buried 400 meters beneath Chesapeake Bay, was large enough to affect most of the mid-Atlantic coast, including possibly the creation of the Bay itself.

Meanwhile, a gigantic crater 400 kilometers in diameter may lie beneath the Arctic ice pack, but so far not yet positively identified.

The larger an object is, the less likely it is to strike Earth, but the more devastating is its collision. A graph of the impact rate of known collisions is included in **5.4.**

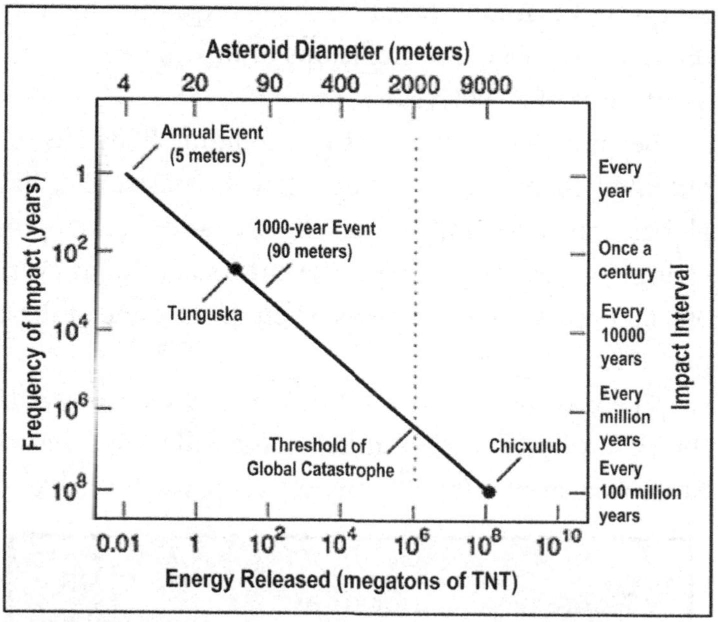

5.4. Earth-bolide impact rate. NASA, *Houston Chronicle.*

Based on current cratering rates an impact of the size that destroyed the dinosaurs occurs about every 100 million years. Smaller collisions, such as the one that left the Barringer crater, happen about every 10,000 years. Earth probably has experienced more than 7,000 impacts of the Barringer size or larger since the end of the dinosaurs. Yet fewer than 200 craters of any size stand confirmed so far. Table **5.5** charts ten of the largest ones geologically confirmed as impact craters. For example, the Manicouagan crater in Quebec, Canada, is in the 100-kilometer (62-mile) class, caused by an object of three to six kilometers (two to four miles) diameter. At more than 200 million years old, craters like this one in the 100-kilometer (62-mile) class occur about every million years. Statistically, there should be 200 more such craters undiscovered on Earth younger than Manicouagan.

46

Doomsday profile

The impact of a bolide between a half kilometer and a kilometer (about 800 yards) in diameter will produce a major disaster such as would occur with an explosion of a very large atomic bomb, enough explosive force to obliterate whole cities. Collisions of this size may occur at intervals of 70,000 years.

Beyond a kilometer in size, a colliding body has the potential to produce a major global catastrophe, like global warming and other climate and life-support changes. These larger impact events occur about every 300,000 years. No one knows when the last one of these occurred on Earth.

There is more. Every 13 million years, an object in the 5-7 kilometer (3-4 mile) class will strike Earth. Moreover, every 100-200 million years we are liable to

5.5. Ten of earth's largest craters. The screened image beneath the table is Vredefort, in South Africa, the largest and oldest known impact crater on Earth The Planetary and Space Science Centre.

	Name	Location	Dia. km	Age mil yrs
1	Vredefort	South Africa	300	2023
2	Chicxulub	Mexico	150	66
3	Sudbury	Canada	130	1849
4	Popigai	Russia	100	35
5	Manicouagan	Canada	100	215
6	Acraman	Australia	90	580
7	Morokweng	South Africa	70	145
8	Kara	Russia	65	70
9	Beaverhead	United States	60	600
10	Tookoonooka	Australia	55	112–133

run into a massive asteroid on the scale of the one that destroyed the dinosaurs and thus forever altered the ecosystems of the planet, or the one that destroyed 90 percent of life in the Permian extinction before that.

Even larger impacts are possible, something about the size of the comet that blew a hole in the side of Jupiter in 1994. The largest fragment of that comet struck with a force 500,000 times greater than the Tunguska blast. A force of such magnitude would mean the end of planet Earth.

The dinosaurs became extinct about 65 million years ago; the Permian extinction occurred 250 million years ago. If the estimate of 150-200 million years between mega-collisions is correct, Earth is not due for another life-altering, catastrophic impact for several million years. Nevertheless, Earth is a dangerous place. Each succeeding generation of inhabitants moves closer to sure annihilation.

Exact calculation of the odds of Earth colliding with something is imprecise because we do not know where all the threats are. NASA estimates that there are about 100 million asteroids. Imagine what the evening sky would look like if they were all illuminated. Very few, however, are reflective enough or large enough to be seen. Out of the 100 million that are out there, 160,000 are in the ten-meter (33-feet) or more class. A thousand belong in the size category of one kilometer (3,281 feet) and above. These 1,000 asteroids are of sizes significant enough to enter the asteroid watch catalogue as potential threats, a half dozen of them known to be larger than 300 kilometers. Another 20 are in the 200 to 300 kilometers range (186 miles). Mapping can account for almost all of the largest asteroids. However, only about half of those in the lesser 10 to 100 kilometers (six to 60 miles) size are on the map so far. These are the size of the object associated with the extinction of the dinosaurs, and mappers know practically nothing about their orbital trajectories.

At least two dozen known asteroids are ten times larger than the one that destroyed the dinosaurs. A collision with any of these objects could destroy the planet. The asteroid Ceres, for example, is more than 30 times bigger than the extinction asteroid. However, these larger asteroids do not pose an immediate danger or a cause for concern.

47

Astronomers calculate the odds of objects from the solar system hitting Earth according to the proportions of them distributed in the different sizes throughout the system. Since there are fewer large objects than small ones, the odds of colliding with a large object figure accordingly.

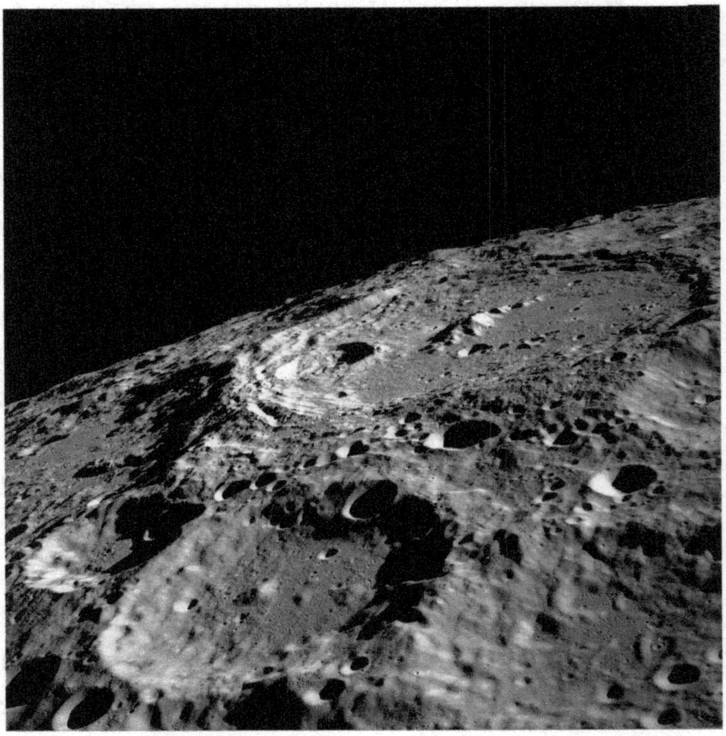

5.6. The Moon's cratered surface. NASA.

Surprisingly, how often we experience an impact — the time span between collisions — is determined primarily by studying the craters of the Moon. A section of the Moon's cratered surface appears in **5.6**. The craters of the Moon range from millimeter size to almost 300 kilometers (186 miles). The Moon's surface shows a large number of craters in all sizes. Unlike Earth, where smaller objects usually disintegrate before striking the surface, the Moon has no atmosphere to protect it.

Because the Moon is in approximately the same location as Earth in the solar system, the frequency of impacts on the Moon in the various size classifications of objects are a good estimate of the time spread of events likely also to be encountered on Earth in a similar period. That is, if the Moon crater evidence suggests only a few large size collisions, and there are only a limited number of objects in the solar system capable of making such craters, then the odds of such an encounter are low. On the other hand, if the Moon craters suggest more frequent impacts in a certain class and there are a large number of things floating around in space that are capable of making such craters, then the odds go up.

48

A program supported by the government and the scientific community exists to identify potentially dangerous objects, that is, those that someday could be on a collision course with Earth. Very little time, however, actually goes into trying to locate new objects or to assess the hazards of known ones; or, for that matter, in planning what we would do if one came at us. No contingencies exist for a plan of action that would save the planet should we discover ourselves to be in the path of a giant asteroid or comet. In addition, even if we had a plan, such as intercepting an incoming object and either destroying it with an atomic device before it reaches Earth or diverting its path somehow away from the planet, scientists agree that the probability of success would be low. In most cases, there would not be time to put together a program and launch a mission to save Earth.

In January 2002, the Near-Earth Asteroid Tracking group recorded an asteroid estimated to be 305 meters (1,000 feet) in diameter. It came within 805,000 kilometers (500,000 miles) of Earth. An asteroid of that size has a potential explosive force of 4,000 megatons, enough to wipe out a large metropolitan area. This object went undetected until two weeks before it flew by at a speed of 105,000 kilometers (68,000 miles) per hour. Experts believe we may experience similar "near-miss" encounters about once a year.

Usually, trackers follow large renegade asteroids over a period of years, and could spot one in all likelihood in time to take interdiction steps if the means existed to do so. We would know a year or two in advance of any giant meteoroids or asteroids whose orbits posed a threat to Earth.

However, not all bolides come from within our solar system. Asteroids do; comets do not. They are yet another problem. Unlike asteroids and meteoroids comets come from beyond the outer planets of the solar system, and they usually appear unexpectedly and move very fast. A meteor travels at 1,800 to 8,400 kilometers (1100-5200 miles) per hour. Comets, on the other hand, approach the Sun at more than 250,000 kilometers (155,000 miles) per hour, or about 70 kilometers (44 miles) per second, more than 50 times the average speed of a meteor.

Any object on a collision course with Earth will come under Earth's gravitational pull. Gravity will cause it to enter the atmosphere at a minimum of 11.2 kilometers (7 miles) per second, 40 times the speed of sound. Any object coming from outside the atmosphere aimed at the center of Earth would take less than three minutes from entry into the atmosphere to point of impact, and most objects are traveling at much higher speeds than this. Impact times would occur in seconds.

Because of such high velocities, the trajectory of a near-earth object becomes even more important. Near misses become easier to understand in estimates of time than distance. For example, in 1989 an asteroid 2.5 kilometers (1.5 miles) across passed within six hours of Earth. More disquieting is the fact that astronomers discovered it *after* it passed.

There may be as many as 100 billion comets of different sizes found in a halo of matter beyond the outer edges of the solar system. Their orbits vary from a few years to several million years. A comet comes into Earth's range about every year. One comes into visible range every ten years. Records exist for some 650 comets scientifically observed.

The largest comet discovered so far was one 68 kilometers (42 miles) in diameter; the smallest detected has been 0.5 kilometers (0.3 miles). Comet nuclei are fragile and break into pieces on many occasions, most significantly in the comet Shoemaker-Levy 9 before it hit Jupiter.

There would not be time to react to a comet on a collision course with Earth. The Lexell comet, for example (discovered in 1770), first appeared on June 14, and by July 1, it was within striking distance of Earth. If such a comet appeared unannounced on a collision path with Earth today, nothing could stop it.

We need not think that such rapid approaches are outdated. In 1996, Comet Hyakutake took just 55 days from the time of discovery until its Earth approach, coming within 15 million kilometers (9.3 million miles) of Earth. Had it been on an Earth collision course, again nothing would prevent its collision with the planet.

49

Near Misses

Many large celestial objects have hit Earth; many more have come close. The 18th century saw an unusually large number of close approaches of comets. Eight comets came within 50 million kilometers (31 million miles). The comet of 1743 came within 18 million kilometers (11 million miles); and the Lexell comet, mentioned above, the closest known approach by a comet on record, passed a mere 2.2 million kilometers (1.3 million miles) from Earth, a near miss in astronomical terms. Coming so close, the Lexell comet must have been a terrifying sight for people living at that time in 1743. Descriptions of it say the comet appeared suddenly and came seemingly from nowhere. At its closest approach, it loomed in the sky almost five times the diameter of the Moon. Astronomers were puzzled about Lexall's rapid approach and erratic orbit until they discovered that the planet Jupiter temporarily locked onto Lexell with its gravitational field, causing a slingshot effect to hurl the comet toward Earth.

For centuries, people thought that craters were not from impacts but rather the remnants of volcanic eruptions, and that impact craters were preposterous science. A pioneer in the decipherment of Earth crater events was the late Eugene Shoemaker. Shoemaker's work with Earth craters did not enjoy universal acceptance until 1994 when the comet that bore his name took its serendipitous course into the planet Jupiter. It was Shoemaker, too, who discovered the 1989 Earth near-miss asteroid mentioned above. Today, Shoemaker's theories on impact craters are accepted science; extraterrestrial objects do strike the planet with calamitous force.

50

Over the billions of years of its existence, Earth has known many narrow escapes. Most of them unrecorded by history but imprinted on the surface of the Moon. Looking at the Moon's near side, the side always facing Earth, it is a sobering thought to realize that many of the huge impact craters, some up to 295 kilometers (183 miles) across, were caused by massive objects that flew past Earth. The objects came close enough to touch Earth's upper atmosphere on their way to collide with the lunar surface.

At the same time, the backside of the Moon, the side out of view, we know from photographs also shows severe scarring by craters. In fact, many more impact craters mark the backside of the Moon than the side facing Earth.

The Moon crater Copernicus (**5.7**) is about 100 kilometers (62 miles) across, and from its pattern of impact debris it appears to have been caused by a near head on collision with the Moon. A relatively young

5.7. Copernicus Crater on the Moon. The crater is seen almost edge-on near the horizon at top center. NASA.

Moon crater, only a few million years old, Copernicus does not align directly with Earth. Still, it represents one of the all-time near misses in Earth's geologic history.

In Figure **5.7**, the large crater centered at the top is Copernicus, compared to one that is 20 kilometers (12.5 miles) across below it. This view came from the orbit of the Apollo 17 lunar expedition; ordinarily Copernicus is oriented more toward Earth.

The large smooth areas on the near surface of the Moon, called *maria* (Italian for seas), are thought to be the result of volcanic changes that erased old Moon craters that may have existed in these lunar regions at one time.

Assuming the same side of the Moon has faced Earth for a significant length of time — say several million years — there should be fewer direct lunar impacts along the axis joining Earth and the Moon. Earth acts as a shield, intercepting incoming objects that would otherwise hit the Moon. Perhaps the lunar seas appear without craters because Earth protected them from hits. Similarly, has the Moon at various times shielded Earth from impact as the Moon's cratered backside may suggest? We have no way of knowing how close some of the Moon collision objects came to Earth on their way to a collision with the Moon, or, on the other hand, what kind of danger the Moon absorbed by its back side before an object could reach Earth.

5.8. Indian rock petroglyph. William C. Miller and Museum of Northern Arizona.

The American Indians may have recorded at least one such near miss. Native American Indians were careful observers of the sky. A thousand years ago, the Pueblo Indians in a pictograph seen at White Mesa, Arizona (**5.8**) recorded what appeared to be a celestial event. A controversy arose over the interpretation of the markings. What kind of celestial event could have occurred in the space between the

Moon and Earth? Stars or planets cannot overlap the crescent Moon. The angle of the Sun's natural illumination pattern renders such an image unlikely as depicted, but possible if the object actually was between Earth and Moon. Such astronomical events were magical to native cultures and well worth recording for their unusual appearances and the omens that they might represent. Still, it is unlikely that an event of such significance to one set of observers that they would immortalize it in stone would appear only on the North American continent. The only other record of a cosmic event that is contemporary with the petroglyph in **5.8** and known to have survived from other world cultures is the supernova of 1054. However, the supernova would not have appeared inside the profile of the Moon. Meantime, other similar petroglyphs in the same region of Arizona do not show the overlapping of images as the one in **5.8**. It could be that one excited artist simply did not observe the event accurately.

Collisions in space are a serious matter. When one considers the Jupiter-Shoemaker-Levy 9 collision of 1994 and its potential damage had it struck Earth, or how the Lexell comet could have altered the planet had it struck in 1770, or any of several of the lunar impact objects, there is reasonable cause for concern about extraterrestrial encounters. Beyond the solar system, we know nothing about future dangers from outside the system. In the vast expanse of space and the millions of years of time that measure the movement of the stars, if another significant object is destined to strike Earth, it is already on its way.

Yet, most experts agree that an Earth defense program to counter such occurrences would not be practical. Besides, analysts say, the likelihood of an apocalyptic ending to Earth coming from outside the planet is less likely at this point than one delivered by our own hand, perhaps ironically as a result of trying in the future to develop a counter bolide atomic weapon. We are already in greater danger statistically of destruction by our nuclear weapons than having them save us. Statistically, we are more likely to blow up the planet ourselves before an extraterrestrial collision annihilates it.

Postscript

Up to this point, distance and scale were in kilometers and equivalent miles. While the English system of feet and miles continues in use extensively throughout the world, the international standard of measurement is the metric system. The great distances in the universe beyond the solar system make measurement equivalencies insignificant, therefore only metric measurements henceforth appear.

Louis Félix de La Rue, after François Boucher. *Astronomy*. bef. 1765. Etching, 7 13/16 x 11 1/16 in. (19.9 x 28.1 cm.) Harris Brisbane Dick Fund, 1953. The Metropolitan Museum of Art.

THE CELESTIAL SPHERE

For my part I know nothing with any certainty but the sight of the stars makes me dream.

Vincent Van Gogh.
Artist, 1853-1890.

Stefano Della Bella. *Galileo and Personifications of Astronomy, Perspective and Mathematics*, 'Opere di Galileo Galilei'. 1656. Etching, 8 5/16 in. × 6 in. (21.1 × 15.3 cm.) The Metropolitan Museum of Art.

VII

MAPPING THE HEAVENS

51

Points of light in the sky may be single stars or galaxies. They may be clusters of stars or groups of galaxies that seen with the unaided eye appear to be a single point of light. To the early astronomers these points of light were all stars. Even the planets were special moving stars. The word *planet* in the Greek means "wanderer," a term used to distinguish the five planets known to the Greeks to *wander* through the constellations. The Sun and Moon were also wandering planets in the Greek system of the starry heavens.

Those who find beauty and solace in the stars, mimic the ancient perception. Planets, stars, and galaxies combine for the modern observer in a panorama of light seldom expressed as individual components in their true nature. Stars, planets, and galaxies blanket the sky undifferentiated in kind in the common experience.

The artist Georgia O'Keeffe, in her painting, *The*

6.1. Georgia O'Keeffe. *The Evening Star, III.* 1917. The third of eight variations, Watercolor, 9 x 11 7/8 in. Museum of Modern Art. © 2022 Georgia O'Keeffe Museum / Artists Rights Society (ARS), New York.

Evening Star, III (**6.1**), saw the sky as part of the landscape. "And the Evening Star would come when it was still sunny," she wrote. "It would be still bright daylight and there would be the Evening Star sitting up in the sky, which I thought was very exciting; and I began painting the evening star Venus."

More than once the sheer beauty of the stars eclipsed the wonder of their physical considerations. Walt Whitman captured the lonely grandeur of the night in 1865 when he wrote the poem, 'When I Heard the Learn'd Astronomer'.

When I heard the learn'd astronomer;
When the proofs, the figures, were ranged in columns before me;
When I was shown the charts and diagrams, to add, divide, and measure them;
When I stirring heard the astronomer where he lectured with much applause in the lecture-room,
How soon unaccountable I became tired and sick;
Till rising and gliding out I wander'd off by myself,
In the mystical moist night-air, and from time to time,
Look'd up in perfect silence at the stars.

From before the beginning of history, the sky has witnessed the passing of events, in ways that connect us to our ancient past. On the Salisbury Plain of England is a circle of giant stones that stands as an enduring reminder of early humankind's reverence for the sky and the lengths to which some people went to memorialize the magnificence of the heavens. Known today as Stonehenge, its stone alignments suggest that it was once a giant calendar capable of predicting celestial events, from the seasonal equinoxes to eclipses of the Sun and Moon. In ruins for centuries, no one knows how Stonehenge was built or by whom (**6.2**).

6.2. Stonehenge, Salisbury Plain, England. Photograph by Diego Delso.

Similar solar calendars existed in other parts of the world as well. One made of wooden poles in place of stones served the American Indian Cahokian culture as late as A.D. 1100. Rings of ritual poles were still in use by the Algonquin Indians when the first Europeans arrived in America. These wood constructions were less monumental than Stonehenge, but they suggest that for thousands of years there existed for many people an intuitive blending of astronomy and theology. A natural dependence on Sun, Moon, and stars informed if not governed events on Earth.

Observations of the heavens commencing long before the invention of the telescope have marked the passage of time for eons, as stars, Sun, and Moon moved in repetitious phases and predictable journeys through the seasons. Hardly a thing has changed in this respect since the beginning of human kind. The same stars that once shone over the people of the earliest ages still light our evening sky. The ancient rituals are gone but replaced by modern ones. Those who probe the mysteries of the universe see in it the same magical qualities of grandeur and awe that attracted all generations.

52

The constellations and the Zodiac

Constellations are flat, planographic configurations of observable points of light in the sky. Taurus, Leo, Sagittarius, and Scorpio are ancient names that evoke the mythologies of the stars. The first astronomers used these characters and others like them to map the night sky. Stars and galaxies joined in groups of celestial bodies comprised constellations whose epic astrological and astronomical influences survive in our conscience if not our culture.

In the relatively brief span of civilization, the stars became symbols of continuity in a history of geographic change and cultural evolution. From the mutable imperfections of the terrestrial plane, where wars and cataclysmic events punctuated human progress, to the imagined perfections of an after-life, there has always been the reassurance of the immortal sameness of the stars. If the ancient cartographers of space were able somehow to return, they would readily recognize many of the same constellations that they knew in their time. With only minor changes, the stars appear today to be in about the same place they were thousands of years ago. Moreover, the same stars still shine with the same apparent brightness now as when Hipparchus first catalogued some 850 of them nearly 22 centuries ago.

The earliest civilizations believed, and many people today still believe that the different alignments of the planets within the Zodiac constellations influence if not foretell the life of a person born at a particular conjunction of Sun, planets, and stars.

The Zodiac is a band of 12 constellations spaced around the ecliptic path traveled by the planets in their orbits around the Sun. Each constellation of the Zodiac, according to astrologists, occupies an assigned portion of the calendar year. A station of the Zodiac imprints its characteristics on anyone born under its influence. This idea guided civilizations, including the most enlightened ones, for centuries.

The practice of grouping stars into constellations is very old. Prominent in Greek mythology, the origins of some constellations date at least to the early Egyptian and Mesopotamian cultures. Many of the symbols used for modern identification of the Zodiac and many other constellations are the same ones known to ancient astrologers. An Egyptian representation of early

constellations, for example, is in Figure **6.3**. Circular Egyptian Zodiac calendars, such as the one pictured from the 1st century B.C.E., show the constellations arranged in an order similar to what we see today. Easily discernable are Leo, Taurus, Pices, Aries, and other recognizable Zodiac symbols.

6.3. Egyptian Zodiac of Dendera. From the Chapel of Osiris, Temple at Dendera, Egypt. Approximately 2.5 meters. Musée du Louvre, Paris.

A few constellations may even predate civilization because some of the animals chosen as names for constellations suggest unique geographical origins that are neither Mesopotamian nor Egyptian. Certainly, the history of the constellations dates at least to the recorded birth of civilization. By 3000 B.C.E., the Sumerians had a well-developed system of astral references. Afterwards, many of the same constellations adopted by the Mesopotamians appeared in other cultures.

Notwithstanding the similarities in constellation names, the twelve constellations associated with the modern Zodiac, did not appear until the era of Nebuchadnezzar and the Neo-Babylonians, around the 5th century B.C.E., arriving in Egypt a couple of centuries later via the conquests of Alexander the Great. This twelve-constellation system is essentially the one that survived through the middle Ages until the present time.

6.4. Astrology Clock of Saint Mark's Cathedral Tower, Venice, 15th century. Michael Jörgens.

53

The Zodiac is associated with astrology, more than it is with astronomy (**6.4**). Both words derive from the Greek language. Astrology refers to the lexicon that describes the stars, while astronomy refers to the laws that govern their positions. Astronomy, although it grew separately from astrology, was likely in the beginning one and the same in practice. Therefore, the origins of the constellations trace more readily in the ancient stories of astrology and religion than to astronomy and science.

Astrology appears strongly represented in early Eastern and Middle Eastern cultures. Yet many believe that a form of astrology existed in Egypt even before it appeared in the Middle Eastern civilizations. The constellation Leo, for example, may go back as much as

9,000 to 11,000 years in Egypt. Based on precession calculations of Earth (precession is the manner in which Earth progresses through the constellations in the ecliptic path), it would have been the astrological Age of Leo about that many years ago — 9,000 to 11,000 years. Some archaeologists believe the Sphinx actually commemorates the Age of Leo. Others are not so sure. No one knows exactly when this half-man, half-lion monument came into existence.

Other striking associations exist between various religious movements and possible astrological ages. It takes the Sun 25,800 years to make one complete precession cycle through the Zodiac, thus it enters a new constellation about every 2,000 years (25,800 divided by 12). Working backward through time, astrologers point to some interesting coincidences. The symbol of Christianity, the fish, for instance, is the symbol of the Zodiac Pisces. The Age of Pisces is the age we are currently in from zero to about 2150, which roughly coincides with the Christian era. Around four millennia ago, the ram, symbol of Judaism, coincided with the Age of Aries the ram, which was roughly 2150 B.C.E. to zero; and before that, the Age of Taurus the bull links to the pre-Biblical Earth cults of Baal, and so on. A handful of conservative theologians make the case even for the Age of Gemini, from about 6900 to 4700 B.C.E., as being a reference in ancient Zodiacs to the twins Adam and Eve.

Interesting as all this may be, the use of astrology as a dating technique is not widely accepted among archaeologists. It is too imprecise and open to interpretation. For example, many believe we are no longer in the Age of Pisces but left Pisces in the year 2001 when we entered the new Age of Aquarius.

Nevertheless, astrology does point to a very ancient origination of the constellations. There are no historical records of the beginning of the constellations, so we will never know for sure their true origins. Over the years, astrologists added many constellations to the original 12 of the ancient Zodiac. Today, eighty-eight recognized constellations map the celestial sky. The Greeks had 48. Other than as a method of grouping sets of stellar objects in two-dimensional proximity, constellations have no scientific meaning or importance. Nevertheless, the ancient astronomer-priests saw no scientific differences between astrology and astronomy. Today, however, there is a distinct division. Astronomers know that the Zodiac constellations do not correlate with the signs of the ancient Zodiac of astrology. Instead, the signs of the Zodiac are a loose reference to 12 equal divisions of the ecliptic path, each named for the original constellation that happened to lie closest to its respective division. The constellations themselves do not center on the divisions of the Zodiac, as many people believe. The whole idea that astrology links to the precise placement of the stars is mythological. Nevertheless, many public libraries contain more works on constellations and astrology than astronomy, attesting in some measure to the ongoing popularity of the magic of the stars.

54

Constellations in the third dimension

Appearing from Earth as proximately grouped stars, constellations are illusory. The bodies in a constellation are much farther apart than the visual grouping suggests. For example, there are four easily discernible bodies in the Andromeda Constellation. A chart of this constellation is in **6.5**. Andromeda comprises more than four bodies, but the four brightest objects in it are the three stars, Alpheratz, Mirach, and Almach, along with the Andromeda Galaxy, from which the constellation takes its name. Astronomers know Andromeda as Galaxy M31, the 31st celestial object listed in the Messier catalog and one of several hundred objects identified by the French astronomer Charles Messier in the late 18th

century, hence the common astronomical reference to Andromeda as Messier 31, or M31.

Seen clearly in the autumn and winter skies of the northern hemisphere just south of the Milky Way and north of the ecliptic path, a line drawn to connect the two stars in the face of the Big Dipper and continuing

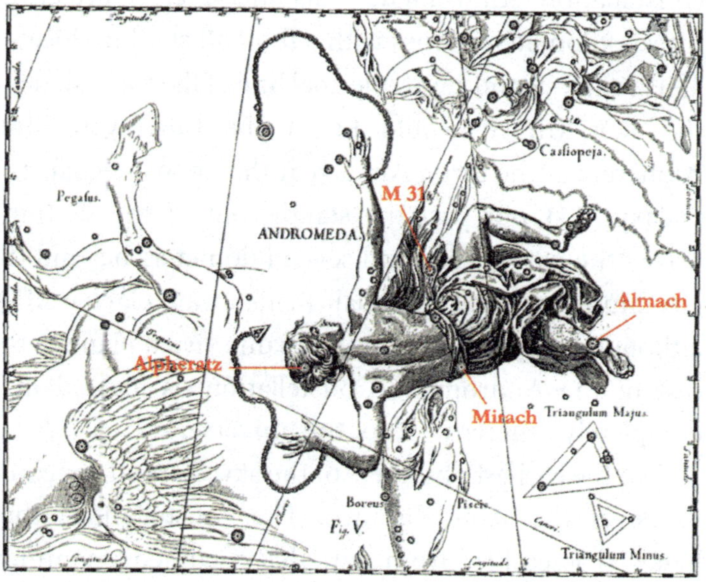

through the North Star will point to Andromeda. This alignment is one way of locating M31 in the fall months. The line will point approximately at the galaxy.

6.5. Constellation Andromeda. Johannes Hevelius' Andromeda from *Uranographia*. 1690. Hevelius drew his charts backwards as if looking in from outside the celestial dome.

The Andromeda Constellation appears to represent a group of closely related stars, as the ancients doubtless intended. In truth, like many constellations, Andromeda contains both stars and galaxies that have no close cosmic distance relationships. The two brightest stars in the Andromeda Constellation, Alpheratz and Mirach, appear approximately four times as bright as M31. This is because Mirach, the closer of the two stars, is more than 22,000 times nearer to Earth than M31. The Andromeda Galaxy, M31, is so far away that if all the other members of the Andromeda Constellation lay at the same distance as M31, then only M31 itself would be visible. All the other members of the constellation would be beyond visible range. However, because it is so far away, M31 actually appears less bright than some of the smaller stars grouped with it in the Andromeda Constellation.

The huge differences in stellar distances within a constellation are difficult to visualize. In the case of the constellation Andromeda, imagine a city measuring fifteen kilometers across with a football stadium located at its center. Standing at one goal line of the football field, the star Mirach would be at the thirty-yard line, Alpheratz at the forty, Almach at the opposite goal, and Galaxy M31 out past the stands, out of the stadium, across the parking lot, and seven kilometers beyond the city limits, or eighty football fields away. Galaxy M31, although it appears to be on the same visual plane as the rest of the Andromeda Constellation, is located in a completely different part of the universe.

To the earliest observers of the sky, stars all appeared to be at the same distance. The constellations they believed assembled from stars that occupied a common space. Yet for what the ancients intended to be sets of closely grouped stars, the constellations are not what they appear to be. The stars and galaxies of a constellation do not lie at all in close proximity.

55

The eye cannot estimate distances beyond a certain point. Our depth of perception is limited. This visual limitation causes far away objects at different depths to appear in most circumstances to be at the same distance from the viewer. Objects at great distances such as stars and galaxies all appear to be in the same plane of vision no matter how far away they actually may be. Near and far bodies look to have the same depth of placement. For the most part, only their relative brightness distinguishes them.

Consequently, two objects of equal brightness, but at different distances in the same line of sight, will result in the more distant one appearing to be less bright than the nearer one although the visual process perceives both objects to be the same distance away.

On Earth, we can usually tell if an object is near or far away based mostly on its apparent size, but we do not know how much farther away one object is from another. In extraterrestrial space, however, size alone tells us nothing about the relative distance of an object. Only brightness is of any use in distinguishing line of sight distances. An ancient astronomical device for coping with this visual phenomenon was the celestial sphere.

56

The celestial sphere

The celestial sphere is an imaginary sphere with Earth at its center onto which we perceive the points of light of the cosmos to be projected. The concept of the celestial sphere is very old. A 13th century representation of a celestial sphere that includes the Zodiac is in **6.6**. The sphere includes the North Star along with an imagined south star. Medieval theologians firmly attached the cosmos to the hand of God. So convincing was the appearance of the celestial sphere to ancient astronomers that they believed it was a physical reality. They thought stars to be the result of a constant sea of light outside the celestial sphere, shining through holes in the sphere. Despite its mystic beginnings, modern astronomers today use the same spherical concept to map the sky and to compile star catalogs, wherein references to the universe rest entirely on the ancient celestial sphere.

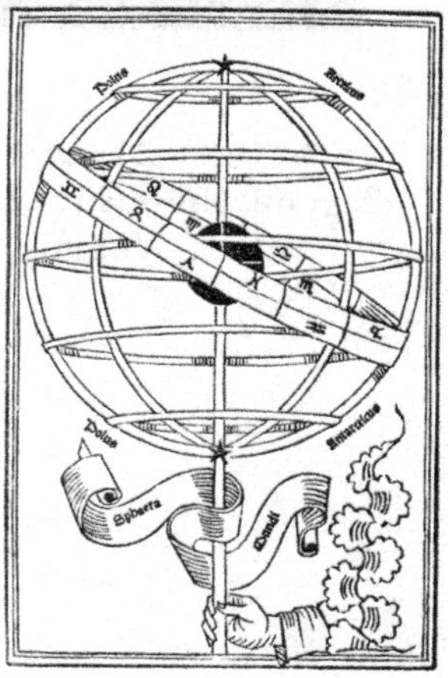

6.6. The medieval celestial sphere. Earth is at the center of the universe in this 13th century woodcut surrounded by a grid, to which the stars were thought to be attached. The Metropolitan Museum of Art.

The celestial sphere divides in an astronomical system similar to the geographical globe system of latitude and longitude. It is the standard reference in mapping the universe relative to Earth, and for specifying the location of astronomical objects in the sky.

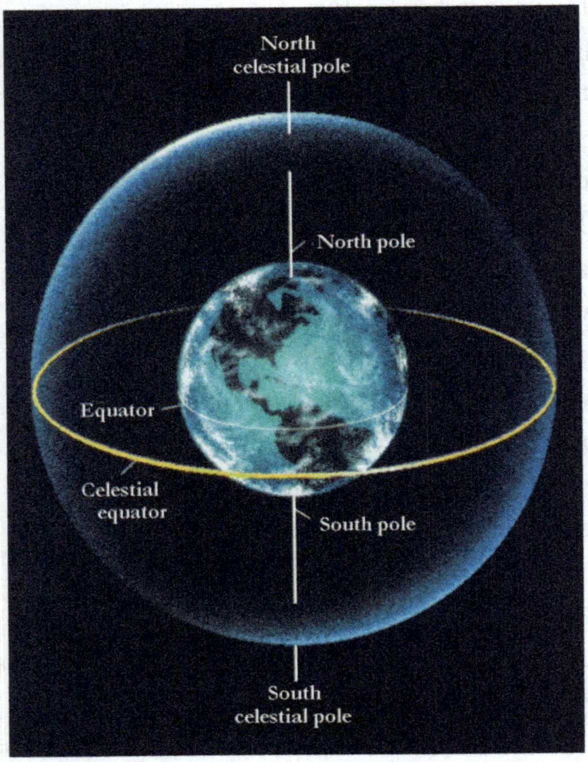

6.7. Celestial sphere diagram. University of Virginia.

The location of an object on the celestial sphere, however, does not describe the depth of the object from the viewer. It has many of the same planographic shortcomings as a map of the Zodiac

The celestial sphere is oriented with the north and south poles of Earth as its axis. The plane of the equator of Earth coincides with the equator of the celestial sphere. The celestial sphere, therefore, is like Earth, only seen from the inside. The equator of the celestial sphere divides into 360 degrees like a compass, beginning first at the celestial prime meridian and then proceeding counter clockwise (**6.7**). The celestial meridian, or zero degree

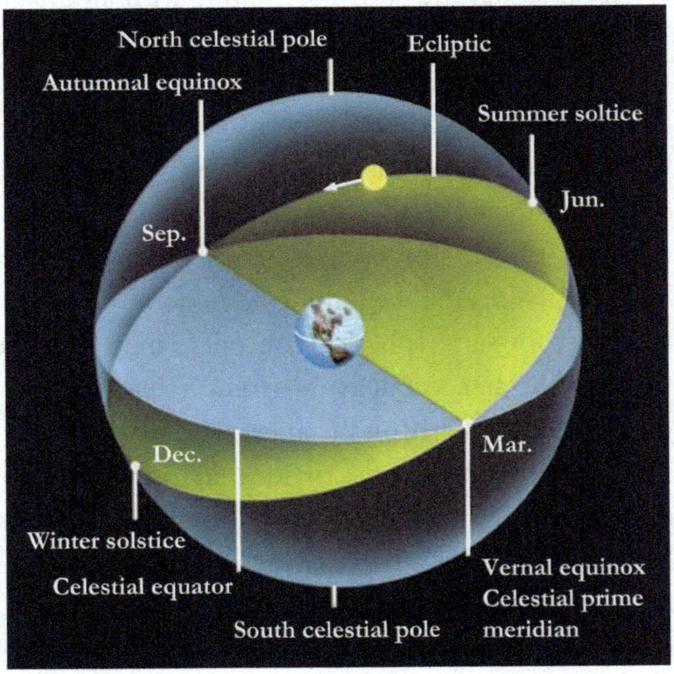

6.8. The celestial meridian. The Ecliptic shows the apparent path of the Sun relative to the plane of the equator. University of Virginia.

point, is the point at which the path of the Sun crosses the equator of Earth, or celestial plane, at the vernal equinox on March 21 (**6.8**). There are two equinox points; each marks a unique time at which day and night on Earth is of equal length, March 21 and September 23. March 21 has the distinction of being the celestial meridian.

The ancients believed in a fixed Earth around which the stars revolved attached to a celestial sphere. From circa 360 B.C.E., when the ancient Greek astronomer Eudoxus first modeled the geometric concept of the wandering of the planets relative to the fixed stars, the celestial sphere has been the model reference for the universe. The celestial sphere was the basis of the Ptolemaic system that persisted for more than 1,500

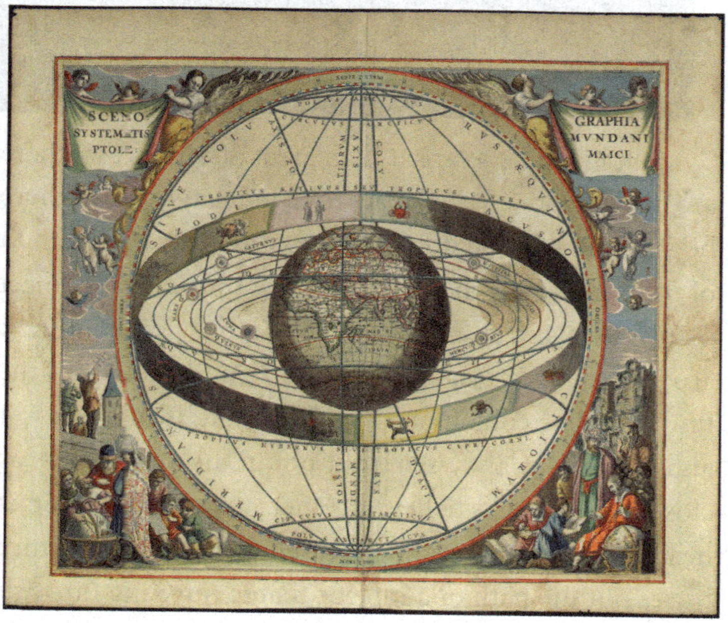

6.9. A 17th century chart showing signs of the Zodiac and the solar system with the world at the center of a celestial sphere. Andreas Cellarius, *Harmonia Macrocosmica*, 1660/61. National Library of Australia.

years (**6.9**), a system that assumed the Sun revolved around Earth.

The celestial sphere is not just a nighttime phenomenon. Under a daytime sky, clouds at high altitudes also appear to float inside a great celestial dome. Farther out still, beyond the atmosphere, we perceive the

Moon in the descending dusk of daylight somehow suspended inside the same dome as the clouds. As the blackness of night compresses the visual layers into a single spherical reference, it is impossible to discern what is inside the dome and what is not. Moon, planets, stars, and galaxies appear to be part of the same great panorama of light.

6.10 Celestial layers. After Juan Carlos Casado.

Celestial objects are at such great distances that there are no perceptual visual clues to their locations anywhere beyond the celestial dome. In the composite photograph of the evening sky in Figure **6.10**, the most distant object seen is a galaxy at left-center. The other points of light are stars or galaxies, also lying at great distances in the universe. The comet, on the other hand, is much closer and is in the solar system (but not in the atmosphere as close as the Greeks believed comets to be). The Moon is nearer still. The other thin line of light across the picture traces out the path of a meteor that is traveling through Earth's atmosphere. Despite great differences in their distances, each of these objects appears to be approximately on the same visual plane within the celestial dome.

57

The spherical interpretation of space is a natural extension of terrestrial experiences. The illusion of a great celestial dome extends perceptually from the natural experience of the dome created by Earth's atmosphere. The atmosphere is a sphere, actual and not imaginary, whose composition of gases defines the visual field to a height of about 80 kilometers (50 miles). The blueness of the sky owes its color to the atmosphere's composition and its interaction with light from the Sun. Clouds and other atmospheric disturbances such as thunderstorms and lightening conform to the lower part of the atmosphere, conditioning our perception of the sky as blue and dome-like. We can sense the atmospheric dome by looking at any point visible in the hemisphere. Everything else in the field of vision appears to curve away from this fixed point toward Earth's horizon. It is easy to imagine how the ancient observers of the sky believed that the stars were in layers in a spherical shell. The experience of the celestial dome easily extends from Earth's atmosphere to a clear night sky of deep space.

Looking up at the enormity of the sky, one quickly understands that any prediction of distance based solely on visual estimates is not possible. Yet, one knows instinctively also that there is information in the varying brightness and sizes of the points of light.

58

The north constellations

In the constellation Ursa Major, the Great Bear, is a group of stars known collectively as the Big Dipper. The Dipper is easily located in the northern sky. It is identifiable by its distinctive dipper shape. The two stars of its face create an alignment that points to Polaris the Pole Star, or North Star. The North Star is located in the handle of the Little Dipper in the constellation Ursa Minor, or Little Bear.

Ursa Major is the constellation that never sleeps, so called because it is visible from northern latitudes at all times of the year. It is not one of the Zodiac constellations, yet it is one of the oldest of star groupings. A favorite of Greek and Roman mythologies, the Great Bear Constellation also figured prominently in the legends of North American Indian tribes. Just how a group of stars that show little resemblance to a bear could evolve as a constellation identified the same way as a bear by civilizations on two disconnected continents is a mystery and may suggest that the constellations are much older than previously imagined. A plausible explanation of the North American Great Bear mythology lies in the probability that its story arrived with the first Europeans, passing to the Indians after a few generations of retelling. On the other hand, the more romantic view would have the legend born in America, transported to the East, and returned as a European usurpation of the original Indian invention. Either way the name Ursa Major is very old.

Northern star groups have been for centuries a beacon of guidance and navigation for travelers and explorers. None other than Napoleon himself — the second most written about figure in history — observed about these stars that "among all the sciences astronomy is the one which has rendered the greatest services to reason and to commerce." The most important of these groups was, of course, Polaris, the North Star.

The celestial North Pole did not always center on the star Polaris. Due to the shifting axis of Earth, other civilizations had other pole stars. Every few thousand years a new star comes into position as the North Star. The star Alpha Draconis, known also as Thurban, for example, was the north star of the Egyptians of 2600 B.C.E. when the pyramids were constructed. In the future, Deneb in the constellation, Cygnus the Swan, will move into position as the pole star. As Earth's precession advances, there will be other stars along the way. About every two thousand years, another star will come into position to become a better polar reference than the current Polaris. Twelve thousand years into the future the star Vega in the constellation, Lyra will likely become the pole star.

59

The North Star fixes the north direction of a compass. It also is a reliable reference for setting latitude using the position of the observer relative to the equator. The altitude of the North Star in degrees above the northern horizon is equal to the latitude of the person viewing it. For example, if the angle of view to the North Star is 40 degrees, the viewer is standing at latitude 40. The higher the North Star is in the sky the closer the observer is to the North Pole, and the lower the star, the closer to the equator.

The North Star is a feature unique to the northern hemisphere. There is no similar prominent star like it at the southern pole. However, the Magellanic Clouds, large clusters of stars and galaxies seen as hazy points of light in the southern sky, historically have served as southern maritime navigational references much like the North Star.

The 24-hour rotation of Earth causes the northern constellations to appear to revolve around the North Star. When viewed from the northern latitudes, certain stars will always remain visible and appear to orbit around the North Star, while the North Star itself appears stationary in the sky because it is almost directly in line with the axis of Earth's rotation. In the southern hemisphere, where no star exactly marks the South Pole, the timed exposure in **6.11** shows the place where it would be located if it did exist directly above the south pole of Earth. The rotation of Earth causes stars to sweep out circular lines around the poles. This circumpolar motion of stars in the north turns the Big Dipper into a clock. The Big Dipper (the back of the Great Bear) appears in **6.12** at different positions eight hours apart. Because the Dipper is visible in the north sky from

6.11. Star trails around the south celestial pole. This timed exposure over several hours shows the south polar center. Anglo-Australian Observatory.

higher latitudes at all times of the year, not only can this North Star group give navigational directions and latitude, it can tell time as well. The Big Dipper will appear to make one complete revolution around the North Star every 24 hours. Knowledge of local star time is required to convert the Dipper to local Sun time, or clock time. Nevertheless, it is as accurate as Earth's progress through time itself.

If the Big Dipper is not the choice, then any group of stars in the constellation Cassiopeia will serve the same purpose because Cassiopeia, too, marks time around the Pole Star.

60

Like the Andromeda constellation, the flat projection of the Big Dipper's pattern of seven stars on the celestial sphere provides no clues as to the great depths that actually separate these bodies in space.

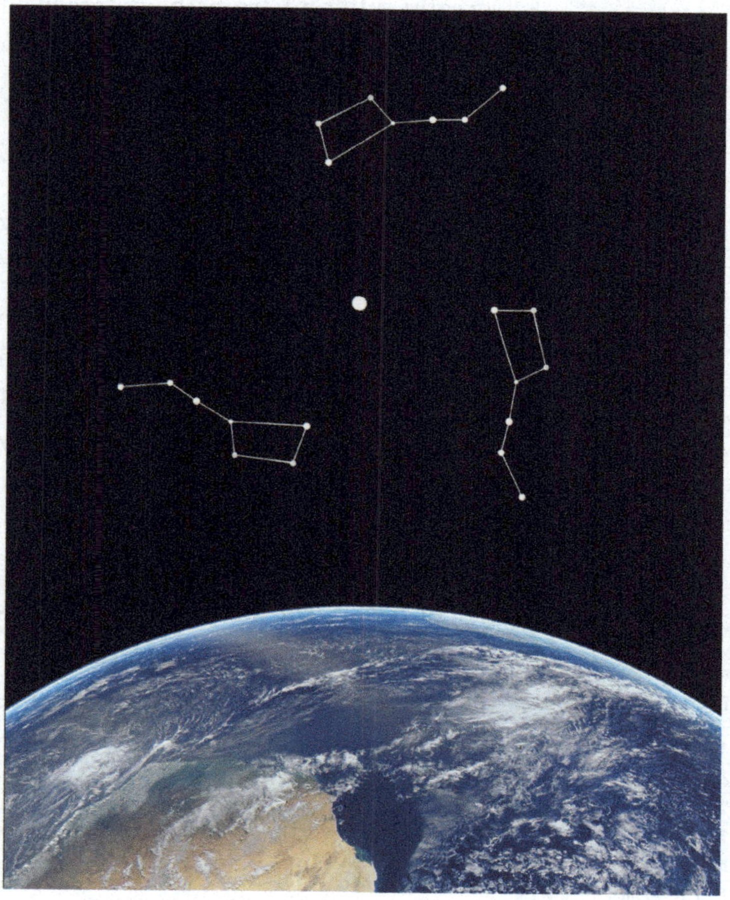

6.12. The Big Dipper clock.

Figure **6.13** illustrates the relative locations of the stars of the Big Dipper from Earth in light years. The size of the point representing each star in the Dipper approximates its absolute brightness in the diagram. Absolute brightness is the estimate of brightness for stars from a common distance beyond the perceived celestial dome. Absolute brightness is how bright the stars would be if they were all at the same distance from the viewer.

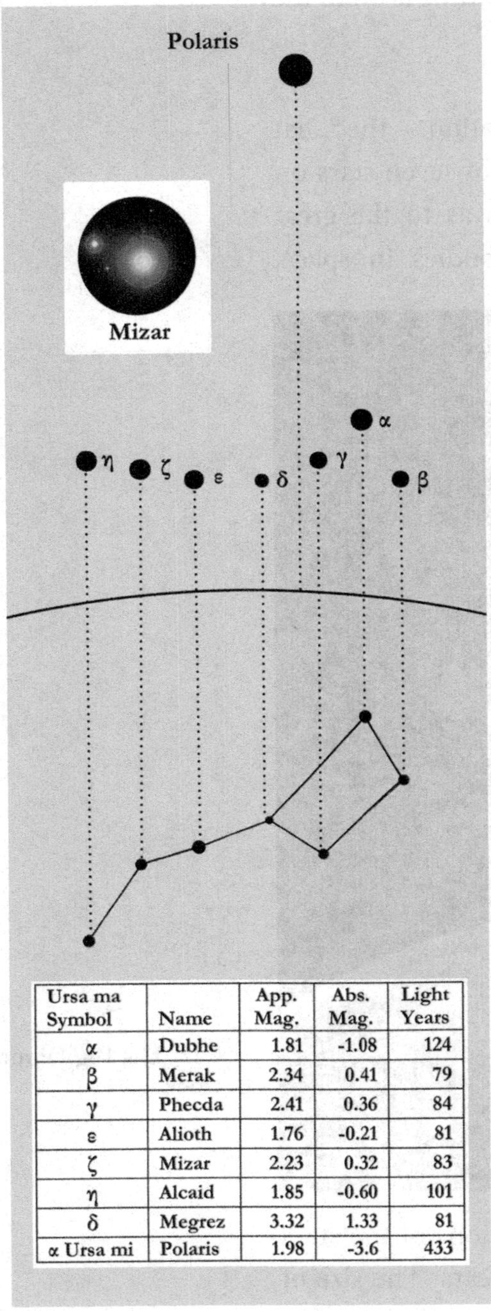

Ursa ma Symbol	Name	App. Mag.	Abs. Mag.	Light Years
α	Dubhe	1.81	-1.08	124
β	Merak	2.34	0.41	79
γ	Phecda	2.41	0.36	84
ε	Alioth	1.76	-0.21	81
ζ	Mizar	2.23	0.32	83
η	Alcaid	1.85	-0.60	101
δ	Megrez	3.32	1.33	81
α Ursa mi	Polaris	1.98	-3.6	433

6.13. The Big Dipper.

On the other hand, relative brightness, or apparent brightness, is what a star looks like from Earth, as it appears on the celestial dome. When true star distances from Earth extend beyond the celestial dome, the actual locations of the stars in the Big Dipper are greatly different from their perceived visual associations.

The Big Dipper appears to the unaided eye composed of seven stars. For most observers this is true. However, the Big Dipper actually contains more than seven stars. Mizar, the middle star in the Dipper's handle, is not one but two stars. Known as a binary star system, Mizar has stood for centuries as a test of visual acuity. Only the best eyesight can distinguish Mizar from its companion star, Alcor. Telescopic examination of the Mizar-Alcor binary shows that Mizar itself is not a single star but is also two stars, Mizar A and B; and there is evidence that each of these stars is in turn a binary system, making of Mizar a quadruple star. Therefore, Mizar-Alcor is a cluster of stars seen as a single star and only detectable by telescopes that can reach beyond the limits of the celestial dome. The Mizar-Alcor quadruple star system appears in the inset in **6.13**.

Other than its distinctive pattern and ease of identification in the sky, there is nothing astronomically unique about the Big Dipper. None of its stars are among the 25 or so brightest in the sky, nor are they among the nearest. Nevertheless, they are interesting for their easily recognized grouping, as well as for their historic role in world exploration.

The brightness of a star is a function of the star's size, type, and distance from the viewer. To the unaided eye, Polaris the North Star appears to be in the same visual plane as the stars of the Big Dipper and to be in about the same apparent brightness category. Yet, in a way reminiscent of the placement of Galaxy M31 in the Andromeda Constellation, Polaris is more than five times more distant than the nearest star of the Big Dipper. If at equal distance with the stars of the Big Dipper, Polaris would appear nearly 100 times more luminous to the eye. The celebrated North Star is also a binary star.

When the great distance to the North Star is located relative to the most distant stars in the Big Dipper, the potential for a vast universe beyond the celestial sphere begins to strain the imagination. When we concede that stars that appear in the same visual plane are substantially more distant than other stars in the same plane, and when we understand that Mizar-Alcor, for example, is not one but at least five stars, each at great distances from the others, appreciation for the cosmos begins to shape a mysterious reality.

Andreas Cellarius. Southern Hemisphere Constellations. ca. 1595. The celestial sphere is seen from outside in a reversed view for a terrestrial observer. Plate from *Harmonia Macrocosmica* of 1661. National Library of Australia.

VIII

THE CENTER OF THE UNIVERSE

61

On any day at any place, we can look about and see where we are. A sense of location tied to the knowledge of community and geography anchors our existence. We may move in our lives — we may even move great distances — but we always retain that unique sense of place that is ours, which we call home.

62

The question that guided much of early astronomy was one of whether the cosmos was heliocentric or geocentric.

Helios was the Greek god of the Sun — the Latin Sol — later identified with Apollo in Roman mythology. Helios was the original sun-centered reference to the center of the universe.

Geo is also from the Greek. The term *ge* is the word for Earth, which when combined with the Greek connective letter *o* identifies anything having to do with Earth. Just as *helio*centric refers to the Sun as the center point; *geo*centric pertains to anything having to do with Earth as the central thing of interest (**7.1**).

7.1. Aristotelian-Ptolemy cosmos with Earth at the center of the universe. Peter Apianus, *Cosmographica*, 1524.

Both the geocentric and heliocentric theories of the universe grew out of separate classical Greek ideas about the motions of the planets. Each theory competed for dominance. Aristotle, for example, believed in the 4th century B.C.E. that Earth and not the Sun was the center of the universe. Aristotle, as it turned out, would of course be wrong.

64

Philosophers, theologians, and astronomers

The classical Greek era was a time of unparalleled intellectual inquiry and philosophical ideas. Alexander of Macedonia — Alexander the Great, a student of Aristotle — extended the Greek empire to the far corners of the known world. Consequently, his expansion of classical Greek civilization spread many of the ideas and theories of cosmic existence based primarily based on the views of Aristotle.

The city of Alexandria was arguably the grandest city of knowledge and culture ever built. It was home to Archimedes, Euclid, Ptolemy, and many others who represented some of the world's greatest thinkers in philosophy, mathematics, and science. Built on the Nile delta of the Egyptian coast and named for Alexander the Great, the boundaries of the city allegedly followed those personally laid out by the young leader. Although he never lived to see Alexandria built, many believe it was the site of his burial.

Alexandria became the intellectual center of the Greek world. The Library of Alexandria — part of the Mouseion, or university, complex — held more than 400,000 volumes, including the writings of Aristotle. According to ancient folklore, no writing existed anywhere else in the world that a copy of it did not exist in Alexandria. Orthodox Christians objected to its pagan teachings and sacked the library in the fourth century. The library, already badly damaged by a previous fire in 269, fell into complete ruin by the time the Muslims arrived in the Arab invasion of 640. With the library's destruction was lost one of the greatest collections of knowledge in the history of the world. Nevertheless, this was not the case for the rest of the city.

Reflecting upon his entry into Alexandria, one of the conquering Arab soldiers wrote, "The Moonlight reflected from the white marble made the city so bright that a tailor could see to thread his needle without a lamp. No one entered the city without having a covering on his eyes to veil him from the glare of the plaster and marble."

In the 2nd century, Claudius Ptolemy wrote his *Almagest* treatise at Alexandria. In it, he placed Earth at the center of the universe according to Aristotle's geocentric theory. The Ptolemy geocentric cosmic system described the motion of the planets relative to Earth. Although eventually proven inaccurate, the system stood as the accepted order of the universe for 15 centuries.

However, not all early Greek theorists subscribed to a geocentric cosmos. The Pythagoreans, for example, who flourished around 500 B.C.E., believed that Earth was not at the center of the universe but was instead like one of the stars, moving with them in circular motion around a great center of fire. The Syracusan Hicetas declared that nothing was in motion except Earth, which revolved on its axis, an idea continued by his alleged pupil, Ecphantus. The Pythagorean, Philolaus — teacher of Plato — went so far as to predict that Earth orbited the Sun in an oblique circle, an ellipse, which obviously turned out to be correct. Plato even appears to have agreed with Philolaus. Meanwhile, Heraclides in the 4th century B.C.E. held the view that the motion of the stars was due to the motion of Earth.

Aristotle remained unconvinced by the arguments of his eminent predecessors, holding instead to the idea that Earth was at the center of a greater universe. He had plenty of support. Geocentric theory dates historically to at least 590 B.C.E. among the Greeks. Moreover,

Egyptian priests allegedly believed in a geocentric universe well before that.

Many of the most respected minds among the early Greeks supported the geocentric idea. The prominent Greek astronomer, Eudoxus (ironically, a student of Pythagorean theory) claimed that the universe revolved in a complex system around Earth. Moreover, Callippus of Syracuse, the most famous astronomer of his time, a follower of Eudoxus, and collaborator with Aristotle, came up with the theory of concentric spheres to describe the apparent motions of the planets and stars. These philosophers and others like them represented the geocentric theory of Aristotle. The Greek Stoic Cleanthes of Assos summed up the prevailing opinion in poetic gratitude to Zeus, when he declared, "The orbs of heaven in circuit round Earth move, and submissive each thy rule obeys." Even those who argued for an Earth in motion stopped short of a position that did not place Earth at the center of existence. Ecphantus said Earth rotated on its axis, but it moved by a divine power that placed it at the center of a cosmological system.

In spite of the sustained popularity of the geocentric theory, the alternative competing theory of a heliocentric system retained credibility. As late as 150 B.C.E., Selencia of Babylon taught heliocentric theory, and Plutarch later seems to have held a similar view, as apparently did the great Roman philosopher and writer, Lucius Annaeus Seneca. Both wrote of an Earth in motion.

Every prominent astronomer, mathematician, and philosopher seemingly had an opinion on this subject. Looking back, Aristarchus of Samos of all the Greeks turned out to be the most accurate of the ancient thinkers. In the 3rd century B.C.E., he clearly understood Earth to revolve around the Sun in a circular orbit with the Sun in the middle. The stars and Sun remained motionless in his system. Archimedes, the great Greek mathematician and contemporary of Aristarchus, and author of 'The Sand Reckoner,' wrote,

"Aristarchus of Samos brought out a book consisting of certain hypotheses, in which the premises led to the conclusion that the universe is many times greater than that now so called. His hypotheses are that the fixed stars and the Sun remain motionless, that Earth revolves about the Sun in the circumference of a circle, the Sun lying in the middle of the orbit, and that the sphere of the fixed stars, situated about the same center as the Sun, is so great that the circle in which he supposes Earth to revolve bears such a proportion to the distance of the fixed stars as the center of the sphere bears to its surface."

Nevertheless, it was Aristotle's views and not the views of Aristarchus that played the most formative role in defining the order of the universe. Ptolemy's theory of a geocentric universe, one in which the Sun and planets orbited Earth, mirrored thinking based on the plan of Aristotle. In opposition to Aristarchus, Ptolemy argued that it would be impossible for Earth to orbit the Sun, as Aristarchus said, because the velocity required for such an orbit would cause the planet to fly apart; thus, he concluded that the Sun must therefore orbit Earth. According to Ptolemy's theory, the Sun, planets, and stars followed perfectly circular orbits around Earth. He introduced a complex set of epicycles to reconcile his belief with the observed elliptical orbits of the planets. Some of the planets, he thought, orbited an unseen point that was itself orbiting Earth.

It was not until the mid-16th century that anyone successfully challenged the long-prevailing Aristotelian-Ptolemaic system. The Polish mathematician Nicolaus Copernicus described the motions of the planets relative to the Sun with such compelling logic and mathematical proof as to establish conclusively the heliocentric idea.

Figure **7.2** diagrams the Copernican system in a print of his original drawing of 1543.

Nevertheless, celestial maps continued to show the old Aristotelian-Ptolemaic geocentric system as the most plausible idea well into the 17th century, considered by many astronomers as a viable alternative to the new heliocentric system. Figure (**7.3**) is a page from the 1660

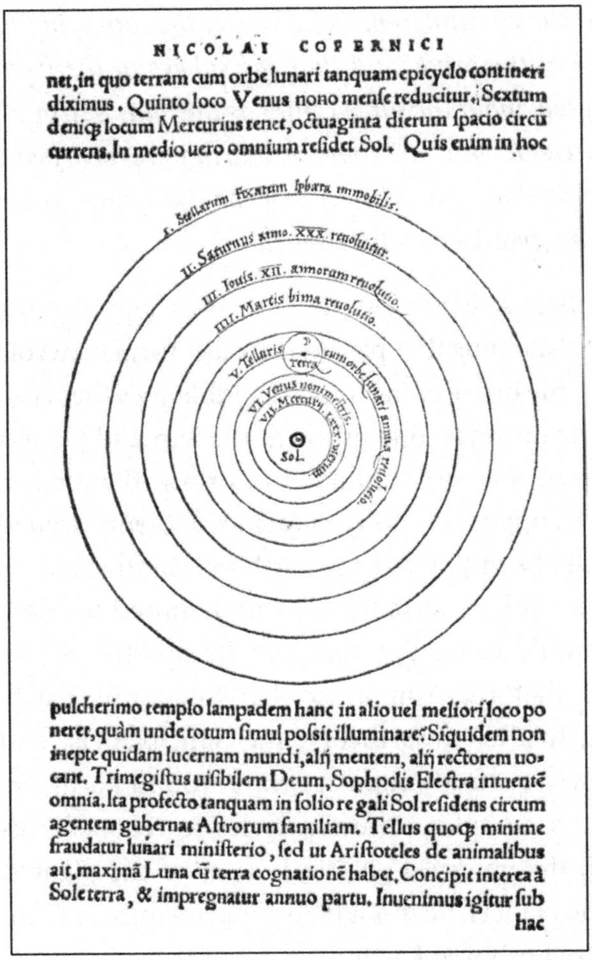

NICOLAI COPERNICI

net,in quo terram cum orbe lunari tanquam epicyclo contineri diximus. Quinto loco Venus nono menſe reducitur. Sextum denique locum Mercurius tenet, octuaginta dierum ſpacio circũ currens. In medio uero omnium reſidet Sol. Quis enim in hoc

pulcherimo templo lampadem hanc in alio uel meliori loco poneret, quàm unde totum ſimul poſsit illuminare? Siquidem non inepte quidam lucernam mundi, alij mentem, alij rectorem uocant. Trimegiſtus uiſibilem Deum, Sophoclis Electra intuentẽ omnia. Ita profecto tanquam in ſolio regali Sol reſidens circum agentem gubernat Aſtrorum familiam. Tellus quoque minime fraudatur lunari miniſterio, ſed ut Ariſtoteles de animalibus ait, maximã Luna cũ terra cognationẽ habet. Concipit interea à Sole terra, & impregnatur annuo partu. Inuenimus igitur ſub hac

7.2. The Copernican heliocentric system. Copernicus' original drawing of the universe published in 1543. *De Revolutionibus Orbium Coelestium* [On the Revolutions of the Heavenly Spheres].

Cellarius *Atlas* published well after Copernicus proved the geocentric theory was wrong.

Copernicus explained his new system with engaging clarity. Among his refutations of geocentric theory, he dismissed Ptolemy's old argument that Earth would disintegrate if it orbited the Sun at high speed by asking, if this were the case, why then would not the Sun likewise come apart in its orbit around Earth. One wonders why

it took more than fourteen-hundred years for someone to point out the incongruity of Ptolemy's logic.

7.3. The Aristotelian-Ptolemaic geocentric solar system, surrounded by fixed stars, with the elements earth, water, air, and fire in the center. The two characters below probably depict Ptolemy and Aristotle pictured in European dress. Andreas Cellarius, *Atlas Coelestis, seu Harmonica Macrocosmica*, 1660.

Questions about the nature of the universe were on the mind of the young Italian artist, Raphael, when he painted *The School of Athens*, in 1510-1511, one of the most celebrated masterpieces of Western art. *The School of Athens* was a product of Raphael's imagination and well-informed research that depicts the competing philosophical ideas at the time of the Renaissance (**7.4**). It is clear from the figures painted in the fresco that in Raphael's time, too, the center of the cosmos was still very much in question. Ptolemy, Zoroaster, and a young Euclid stand and kneel in the grouping at lower right. Ptolemy (with his back to the viewer) holds an Earth sphere. Zoroaster, the prophet of ancient Persia, holds a celestial sphere. Meanwhile, Pythagoras kneels in the foreground at left, while Heraclides — shown in the painting as Raphael's contemporary idol, Michelangelo — reclines thoughtfully on a rectangular writing stand at front center. Some scholars insist that this figure does not represent Heraclides, the 4th century B.C.E. advocate of an Earth in motion, but rather Heraclitus, the 5th century B.C.E. Pre-Socratic philosopher of pantheistic materialism.

Meanwhile, Epicurus who lived around 300 B.C.E. is the second figure in from the left base of the arch. Epicurus first guessed that the basic material of existence is atoms, bits of matter flying through space. Parmenides (standing in the foreground gesturing to the seated Pythagoras) and Zeno, his disciple (left base of the arch facing Epicurus), are both included in the fresco for their

7.4. Raphael. ***The School of Athens.*** **1510-1511. Fresco, 26 x 18 ft. Stanza della Segnatura, Vatican, Rome.**

theories of existence and multiplicity that today still challenge the realities of the universe. Included in *The School of Athens*, of course, are Plato and Aristotle, the two principal figures at the center of the painting. Socrates, meanwhile, stands to the left of center (in yellow). The sparsely clad figure sprawled on the steps is Diogenes, the archetype of cynicism.

Obviously, not all these men lived at the same time. Several centuries separate the oldest from the youngest, including Raphael. A self-portrait of the artist himself appears at the far right of the painting, the second figure from the right base of the arch.

Some of the early philosophers appear to have been remarkably accurate in their predictions about the solar system and universe. Aristarchus, for example, who described a heliocentric system, was correct. Others turned out to be less accurate. The ancient theories are subject, of course, to evaluation in light of modern thinking. Without the full benefit of the original context of the ideas, the reading of the past is subject to incomplete if not inaccurate historical interpretation. For instance, Ptolemy's belief that Earth was at the center of the universe came from spiritual considerations as much as science. After all, Aristotle, Ptolemy's idol, was one of the most dominating thinkers in history to espouse the dual existence of humankind. It was not until the grip of Aristotelian logic began to loosen in the 16th century that the ancient context of meaning began to change from one of religious and philosophical roots more toward science and exploration. Civilization embarked on a renaissance of new ideas that led, among other things, to the establishment by Copernicus that the solar system is indeed a heliocentric system, and thus laid to rest forever the false notion that Earth is at the center of the universe.

65

Today, conventional observation concludes that the universe has no center. Whether once it actually did have one, is unknown. If the Big Bang theory is true, then there must have been in the beginning a center of the universe and one may still exist somewhere, but no one knows where. If there was a center at which it all began, we have lost track of it. Therefore, in the absence of a real center, we imagine one.

A human-centered universe is a universe in which humankind places itself at the center of existence. Such a centric interpretation of being is both theological and scientific. This is one reason why it was necessary for centuries for everything to revolve around Earth. For man to be at the center of the universe, Earth on which humanity resided had to be at the center. However, the dichotomy between astronomy and religion — born of Copernican theory — has grown wider with each new discovery about the nature of the universe. From a time when it seemed impossible to distinguish a physical heaven from its spiritual counterpart, the growing knowledge of the universe has exploded the myths that in the beginning painted humankind as the ultimate outcome of a unique plan.

Gone are the myths, but not the expectations. Science frequently uncovers reasons to question the tenants of a divine human purpose. Yet, so far, there is no direct scientific contradiction of the idea that life on Earth is somehow central to the cosmic scheme of things. If, however, life exists in other parts of the universe, as many expect it does, then the significance of humanity will need reexamination, and the notion of providence at least expanded to a wider realm than to just people on Earth. It will be interesting to see what new turns centric theory takes when new discoveries may erase all physical possibilities of a special role for the human species.

Since the arrival of space travel, there has been an increasing awareness that we earthlings probably are not as significant as the ancient philosophers previously hoped. Ever since we first saw Earth floating in space as a blue sphere, isolated against the blackness of an empty void, surrounded by a thin blue halo of life-supporting atmosphere, the question of existence has become ever more complicated than it was in the days of Aristotle.

7.5. The planets in the solar system, approximately to scale. The planets are, from left, Mercury, Venus, Earth, Mars, Jupiter, Saturn, Uranus, Neptune, and the dwarf planet Pluto. Lunar and Planetary Institute.

A longing to know about the origin of humanity is a common theme running through history. Parallel to this is a universal interest in a sense of place, seeking not only answers to the questions of who we are and how we got

here, but also *where* we are. It is disquieting when confronted with the idea that there is no central cosmic point by which all existence ultimately obtains its description. In the absence of a physical center, each civilization has turned inward to describe itself relative to a spiritual center because unlike the physical universe, the spiritual one requires no proof.

66

The solar system

Earth is one of nine planets — eight if the dwarf planet Pluto is discounted — in the solar system. Earth is the third planet in order from the Sun. Of mid-size among the planets, Earth is larger than four of them are, and smaller than four. Figure **7.5** shows the sizes of the planets relative to the Sun. The equatorial diameters of all the planets arranged from largest to smallest fit comfortably within the diameter of the solar disk. They could line up three times across the Sun with room to spare. Relative to the size of the Sun, Earth is very small. Earth has less than .0000008 percent of the volume of the Sun. The planets are so small compared to the Sun that the combined volume of all of them put together would not occupy 1 percent of the space taken up by the Sun.

When placed in their orbits around the Sun in the vacuum of space, the planets seem even smaller. For example, imagine the Sun as a seven-story building. At the same scale, Earth is the size of a golf ball forty miles away from the building. Figure **7.6** diagrams the orbits of the planets. The outermost solar system member,

7.6. The orbits of the planets in the solar system.

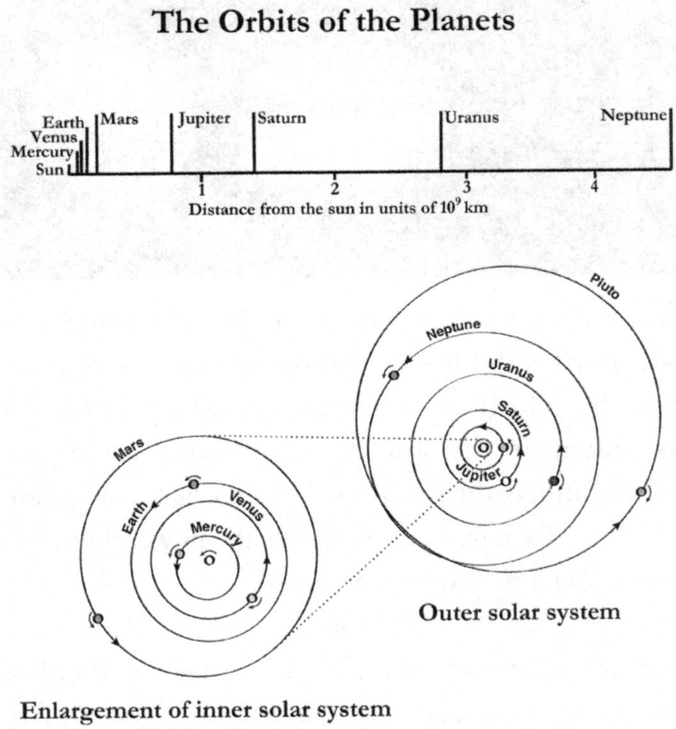

Pluto — the most eccentric of the nine planetary bodies (and last to be discovered in 1930) — traces a path that is on average 12 billion kilometers across, thus inscribing the boundaries of the known solar system.

Each of the planets is unique in its orbit. For instance, from 1979 to 1999 Pluto was at a point in its orbit when it was inside the orbit of Neptune, making Neptune for a while the most-distant planet. Pluto takes 248.5 Earth years to make one revolution around the Sun. Since the birth of the American Republic in 1776, Pluto will not complete one trip around the Sun until the year 2024. In contrast, the innermost planet, Mercury, circles the Sun once about every 88-earth days.

The solar system is relatively young when compared to the rest of the universe. The Sun is 4.5 to 5.5 billion years old. Assuming the universe to be about 14 billion years old, the Sun and Earth were born at about year 10 billion in the life of the universe.

Meanwhile, humankind in the form of *Homo habilis*, the earliest form of humanity, entered the record at about year 13.998 billion. *Homo sapiens,* or modern humans, appeared around 13.9998 on the scale of 14 billion years. The entire history of civilization from about 5,000 B.C.E. until now would not even register its mark on the cosmic time line.

The oldest stars in our galaxy, the Milky Way Galaxy, are 10 to 12 billion years old, which means that Earth came into existence about midway in the formation of the galaxy, and the galaxy was formed comparatively early in the growth of the universe. The Sun is one of more than 100 billion stars in the Milky Way Galaxy, and yet the only star known to possess a solar system such as ours, and the only one so far to sport a living planet.

67

The beginning of modern astronomy

Tycho Brahe was arguably the greatest astronomer to live prior to the invention of the telescope. However, science celebrates his young assistant, Johannes Kepler, who bridged the old astronomy and the modern, as the greatest astronomer of all time. The field of physics remembers Kepler primarily for his laws of planetary motion, which he deduced from the observations of his mentor, Tycho Brahe.

Born near Weil der Stadt, Germany, Kepler as a young man seriously considered becoming a priest, having studied at the seminary (7.7). Instead, following his completion of study at the University of Tubingen, he took a job teaching mathematics. At Tubingen, he became interested in the heliocentric theory of Nicolas Copernicus, and in 1596 at the age of 24, he wrote and published *Mysterium Cosmographicum* [The Cosmographic Mystery], the first effective defense of the Copernican theory in more than 50 years. So impressed was Tycho Brahe with the young Kepler's model explaining the relative distances of the planets from the Sun, that he, Brahe, invited Kepler to Prague to become his assistant — although Tycho himself was not a proponent of Copernican theory. Tycho believed that the Sun revolved around Earth and the planets revolved around the Sun. Kepler, a devout Lutheran tormented by the upheavals of religious reform in Europe at that time, was among Protestants forced to convert to Catholicism as part of the Counter Reformation movement, or leave the province. Therefore, in 1600 he gladly left Germany to accept Tycho's invitation to come to Prague.

7.7. Portrait of Johannes Kepler. Unidentified painter. Copy of a lost original from 1610 in the Benedictine monastery in Kremsmünster.

Within a year of Kepler's arrival in Prague, Tycho died. In 1601, Kepler inherited not only Tycho's position as Imperial Mathematician but also Tycho's papers, a lifetime of the most accurate astronomical observations produced up to that time. From Tycho's records, Kepler showed that the planets move in elliptical paths about the Sun. He showed further that if you draw a line from the Sun to a planet, that line sweeps out an area in a given time equal to any other area swept out in the same amount of time, wherever the planet is in its orbit. This means that when a planet is close to the Sun it must travel faster in its orbit than when it is far from the Sun. These conclusions constitute the first two of Kepler's three laws of planetary motion. The third law relates the time it takes a planet to complete one orbit around the Sun — one planetary year — relative to the average distance of the planet from the Sun. The length of a planet's year squared is equal to the distance of the planet from the Sun cubed. Therefore, the distance of any planet from the Sun calculates simply by knowing how long it takes it to complete one orbit. Kepler's three laws formed the foundation for the later work of Isaac Newton and the formulation of the universal laws of motion and gravity.

While at Prague Kepler wrote important treatises on optics and the telescope. He gave the first modern explanation of the visual process, and wrote and published an open letter in support of Galileo, at a time when the latter was the target of religious persecution.

However, Kepler was never free for very long at a time from religious and political turmoil himself, nor from personal tragedy. Under political duress, he left Prague in 1612, the same year his first wife died. He soon remarried, but of the 11 children produced by the two unions, only five lived past the age of seven. Despite his fame, Kepler and his family did not escape personal persecution. Inquisitors accused his mother of being a witch and jailed her but released her a few years later. This period in Kepler's life also saw the beginning of the devastation of Germany and Austria from the Thirty Years War.

After Prague, Kepler settled in the city of Linz, Austria, where he continued to publish controversial ideas. In 1613/14 he showed that the Christian calendar was in error, and that Christ was born in four B.C.E., today an accepted conclusion (although many believe six B.C.E.). Shortly thereafter, he published *Epitome Astronomiae Copernicanae* [Epitome of Copernican Astronomy], the most influential textbook of the time.

Counter Reformation pressures worsened in Austria. Authorities kept soldiers at one time garrisoned in Kepler's house. He and his family fled in 1626. Still, amid the turmoil he managed to continue his work. Having to recreate much of a previously destroyed text, he was able in 1627 finally to publish his *Tabulae Rudolphinae* [Rudolphine Tables] based on his and Tycho Brahe's observations of planetary motions. History credits these calculations with the conclusive establishment of heliocentric astronomy, that is, that the planets unquestionably do orbit the Sun instead of the Sun orbiting Earth as had been previously taught.

In the end, Kepler found himself without a job and no salary. He unsuccessfully sought appointments from different ruling courts and returned briefly to Prague to try to extract salary there that the Prague court owed him from his years as Imperial Mathematician. He died in Regensburg, Germany, in 1630.

In addition to his major publications, for which history remembers him best, Kepler produced several other works on a host of subjects, including what may be one of the first science fiction tales. Published posthumously in 1634, *Somnium* [Dream] is the imagined account of a man's journey to the Moon and the Moon's inhabitants found there, a tale, incidentally, that predated H. G. Wells' story of the same theme in 1901 by almost three centuries. Always surrounded to some extent by an aura of mysticism, Kepler's religious and philosophical temperament never strayed far from his ideas of beauty and harmony in the universe. He once remarked that his entry into mathematics and astronomy, and his attraction to the Copernican theory of the solar system, came inspired by "metaphysical reasons."

68

The Milky Way Galaxy

The solar system is about 12 billion kilometers across at the orbit of Pluto, the solar system's most distant known member. Yet compared to the diameter of the Milky Way Galaxy, the solar system is a mere speck in the universe next to the scale of the galaxy. Set end to end, one and a half million solar systems would not bridge even the thickness of the galaxy, let alone its diameter. It would require 76 million solar systems to span its diameter. The solar system in the galaxy is like a ten-inch toy boat adrift on all the oceans of the world combined. Earth is like a microscopic dot on the deck of that toy boat

At the center of the Milky Way Galaxy is a nucleus 90 trillion kilometers in diameter, surrounded by a spherical nuclear bulge of 60,000 trillion kilometers. From this central axis, spiral arms twist outward in a flat, disk-like shape 900,000 trillion kilometers in diameter and 18,500 trillion kilometers thick. The thickness of this radial disk is 1/50 of the galaxy diameter, meaning that the galaxy is essentially flat. The center of the galaxy is at a distance from Earth predicted to be 255,000 trillion to 312,000 trillion kilometers. At that great distance, it takes light thousands of years to travel from the center of the galaxy to the solar system.

Consider for a moment that Earth's orbit is only 300 million kilometers across, or one-fortieth of the diameter of Pluto's orbit; and that Earth itself is a mere 12,756 kilometers in diameter. A million Earths set end-to-end would barely cross the solar system; a chain of 12 trillion Earths *in orbit* would not cross the galaxy (**7.8**). Earth — the plane of human existence within the universe — has a scale of 0.0000000000001 of a single average size galaxy. From this, one can easily appreciate the minuscule physical relationship of Earth to the solar system, to the galaxy, and ultimately to the universe. If Earth were the size of a grain of sand, the solar system would be roughly the size of a football field. The same grain of sand in the galaxy would find itself in a void of dimensions beyond comprehension. Imagine a dot the size of a period lost in such a space, and that is the insignificance of the size of Earth compared to the solar

7.8. Artist' rendering of the Milky Way Galaxy showing the approximate location of the Sun. European Space Agency.

system and the galaxy, not to mention our relationship to an unimaginably large universe composed of millions if not billions of galaxies.

69

By the 17th century, the Sun had replaced Earth as the presumed center of the universe. Two hundred years later in the 19th century, it became apparent that neither Earth nor the Sun was at the center, and astronomers began to look for other possibilities.

The Sun, as it turned out, is only one of many similar stars in the Milky Way Galaxy. It is not unique; neither is its position special within the galaxy. The Sun and the solar system are located in the galactic disk between one-half and two-thirds of the way out from the center of the galaxy (**7.8**).

An external view of the galaxy has no one, correct vantage point. One view may show the galaxy rotating counterclockwise while in the other it is revolving clockwise. The apparent reversal of rotation in the two images assumes different perspectives of the galactic disk, one from the top and the other from the bottom.

When we remove Earth and the Sun — geocentric and heliocentric — as candidates for the center of the universe, it is logical to assume then that the galaxy must be the probable center. In fact, the Milky Way Galaxy was for many years up until the early part of the 20th century thought to be the full extent of the universe and that the center of the galaxy, if ever discovered, would be synonymous with the center of the universe.

Long after the establishment of the Sun's common existence among the stars, astronomers continued to imagine that the Sun was special. As late as 1917, astronomers believed that the Sun was near the center of the galaxy. They also thought that the galaxy was a flat disk of cosmic material that extended out only a few thousand trillion kilometers. In the late 18th century, astronomers estimated the size of the galaxy to be no more than 50,000 trillion kilometers, large to be sure but far short of later predictions and nothing of the scale even approaching its known size today. It was only in 1918 that astronomers discovered that the Sun is not at the center of the galaxy; it is but one of billions of stars in a gigantic galaxy of stars whose center is 300,000 trillion kilometers away. At this great distance, radiant energy that began its journey from the galaxy's center before the stone age of humanity has not yet traveled far enough to reach Earth. The energy now coming from the center of the Milky Way Galaxy started at the dawn of prehistory more than 30,000 years ago.

Just as the Sun turned out to be not particularly unique, we now know that not even the Milky Way

Galaxy is special and is almost certainly not at the center of the universe. Since all matter is moving away from the point of the original Big Bang explosion, the origin of the universe could be almost anywhere. It could be between our solar system and the distant edge of space. Perhaps it is behind us; either way puts us between the origin of the universe and its frontiers. The center may be even somewhere beyond the universe's present boundary. Each of these three possibilities for the origin of the Big Bang fulfills the prerequisite of an expanding universe. Cosmic bodies are moving apart at great velocities. So far, they are detected to be moving only relative to each other and not relative to a common point of reference, one that could be interpreted as a physical center, or beginning, of the universe.

IX

CHASING THE CYNOSURE

The earliest civilizations saw the Milky Way as a luminous band of light of mysterious origin, glowing in the nighttime sky. The Mayans of Central America believed the Milky Way to be the mystic pathway that souls walked to the underworld. Hindu mythology identified it as the churning of the milky ocean that marked the eternal struggle between the

forces of good and evil. While in Far Eastern countries, China in particular, the Milky Way was the dividing line between two mythological lovers represented by the stars Vega and Altair whose desire to get together triggered all kinds of events on Earth (**8.1**).

8.1. Milky Way seen over the Chilean Atacama Desert. Photo by Stephane Guisard, European Southern Observatory.

The Greeks had elaborate tales about the creation of the Milky Way, as did the Romans. The word *galaxy* in Greek means "milky way." One story tells how as Juno, wife of Jupiter, lay asleep Jupiter placed Hercules at her breast, and by doing so, he hoped to give Hercules immortality. However, Juno awoke and pushed Hercules away. As she did so, milk from her breast arched across the heavens to create the Milky Way. In an alternate version of this myth, Minerva, the goddess of wisdom and protector of the mortal Hercules, persuaded Juno out of pity to put Hercules to her breast. As Hercules nursed, his aggressiveness caused Juno's immortal milk to spurt across the heavens to create the Milky Way. Some drops fell to Earth where lilies sprouted.

A handful of visionary individuals, like Pythagoras and Democritus, guessed very early that the Milky Way might be composed of individual faint stars, but they had no way of knowing for sure. It was not until 1609 that Galileo turned his newly invented telescope skyward. He discovered for the first time the true nature of the Milky Way, confirming the prophetic words of the 1st century Roman poet Manilius, who wrote, "The spacious band serenely bright; from little stars, which there their beams unite, and make one solid and continuous light." Galileo would eventually reveal many other aspects of the realities of the universe.

71

Galileo Galilei

With the aid of his telescope, Galileo confirmed the Copernican system of planetary motion, exactly as Kepler had defended it. For three years, Galileo did not declare his findings. His discoveries contradicted the accepted teachings of Aristotle that formed much of the doctrine of the Roman Catholic Church. The Church denounced the Copernican system as dangerous to the faith, and warned Galileo. Nevertheless, he published anyway in 1632 a work supporting Copernicus, and not

Aristotle and Ptolemy. His treatise, *Dialogo sopra i due massimi sistemi del mondo* [The Dialogue Concerning the Two Chief World Systems], was particularly inflammatory because he wrote it for the popular, nonscientific reader.

In 1633, the Inquisition tried Galileo and forced him to recant all of his beliefs and writings supporting Copernicus. Legend holds it that as he arose from his knees before the heresy panel, having just denied the truth of Earth's movement around the Sun, he said, "Nevertheless it does move." (**8.2**)

8.2. Portrait of Galileo Galilei. Drawing by Ottavio Leoni, 1624. Biblioteca Marucelliana, Florence.

After a brief time in prison, Galileo lived in seclusion, nearly blind, until his death in 1642. His work remains recognized as a turning point in scientific and philosophic thought on the world and its place in the solar system and universe. Nevertheless, it took until the end of the 20th century for the Roman Catholic Church to rescind its persecution of Galileo. In the late 1990s, the Church finally moved to rectify one of the great errors of human judgment and authority by declaring Galileo innocent of heresy.

72

The cosmic principle

Many of the questions that the ancients asked about the universe still seek answers. The questions are at least as old as civilization.

How did the universe begin?
What exists beyond the universe?
How will the universe end, and when?
What makes up the universe?
Why is there a universe?
Does an omnipotent force control the universe?
How large is the universe?

The boundary of the unknown is well beyond the limits set centuries ago, in Aristotle's day for example. The boundary extends past the known sphere of ancient knowledge. We believe we know more today than was known centuries ago, and doubtless we do.

However, all knowledge of the universe rests on one fundamental assumption. The same laws of physics that operate in our solar system are at work throughout the universe. Space, time, and motion are immutable precepts. Without this cosmic principle, conclusions about the universe are not possible.

73

8.3. Portrait of Isaac Newton. Painting by Godfrey Kneller, 1689. Collection of the Earl of Portsmouth.

Newtonian physics is a term applied to scientific principles based on the concepts and theories of Sir Isaac Newton, and pertaining to, or in accordance with, his work, especially in mechanics and gravitation. Newtonian physics is the basis of the cosmic principle that says that all matter and energy in the universe behave the same as matter and energy observed on Earth. This principle allows astronomers to project the size and shape of the known universe, and to describe its composition. Many of the fundamental ideas of Newtonian physics appeared in Newton's 1687 treatise titled *Philosophiae Naturalis Principia Mathematica* [Mathematical Principles of Natural Philosophy]. History credits him with originating laws on the mechanics of force which are central to understanding the physical world, and which have made space travel possible. His three laws of motion, along with his universal law of gravitation, are fundamental to describing the nature of the universe. The portrait of

Newton in **8.3** shows him as a young man at about the time he formed his ideas on gravitation.

Fellow Englishman Edmund Halley, friend and colleague of Newton, financed Newton's *Principia,* encouraging its publication. We remember Halley best for his namesake Comet Halley. His poem, *Ode to Isaac Newton,* was a clear summary of an emerging understanding of the universe, an understanding attributed to Newton and hailed in the final words of Halley's poem.

Then ye who now on heavenly nectar fare,
Come celebrate with me in song the name
Of Newton, to the Muses dear; for he
Unlocked the hidden treasuries of Truth:
So richly through his mind had Phoebus cast
The radiance of his own divinity.
Nearer the gods no mortal may approach.

Newton's ideas continued into modern times, used by science to explain much of the physical world and the universe. Known also as classical physics, Newtonian physics serves as a general reference to physics developed before the formulation of quantum theory, or relativity theory. We accept Newtonian physics as characterized by objectivity and mathematically precise in description.

However, there are sufficient numbers of exceptions to Newtonian physics to raise reasonable doubt about the validity of all of Newton's ideas as the basis for a cosmic principle. Questions arise, for instance, in the behavior of energy in subatomic structures, in the behavior of light, and in the phenomenon of gravity. Enough questions exist to suggest that such a cosmic principle of mechanics does not apply uniformly throughout the universe.

An alternative way to describe the universe is through quantum physics, introduced in the early part of the 20th century. Some look upon quantum physics and its correlates as merely an addendum to Newtonian physics, others read quantum theory as a very new way of looking at the universe.

74

In the 17th century, scientists discovered calculus to describe mathematically things that are in progress. (Newton claimed to have invented it.) Up until that time, no language existed to describe objects in motion. Calculus was able to calculate processes and make projections. It is so fundamental, modern science without it would be impossible. If, after another millennium of scientific discovery, it turns out that neither Newtonian physics nor quantum physics prove to be universally applicable, a whole new understanding of the universe may be necessary, an understanding of things for which today, as before Newton, there is no calculus, no known language.

75

Perhaps the universe harbors a reality different from anything seen so far. Energy that disappears in one part of the universe, in a black hole, for example, simply may be following a path out of the visible universe to reappear simultaneously in some cosmic event at another point. Energy may exit one universe only to spring up in another in one great burst of energy from a void of such enormous scale that it cannot be associated with human concepts of time and space. Serious theoretical models contemplate just such an occurrence when travel through deep space permits a return to our universe at any point in time, even the distant past. Likewise, space and all that defines it: shape, size, matter, energy, movement, and time, may reduce one day to an idea for which mathematical means of description awaits invention. The option of a mutable universe undermines all possibilities of a logical link between visual experience and reality.

76

A mutable universe

Of all the ancient theorists, Aristotle advanced ideas about the physical nature of things that molded early concepts of the universe. He was among the first to develop systematic ideas about the physical world (**8.4**). Ultimately, his influence shaped the understanding of the universe for more than 17 centuries, and laid the foundation of the philosophical and scientific world for longer than that. While his philosophical ideas remain in high regard, many of his speculations about the physical world were wrong, modified, if not supplanted, by modern theories and discoveries.

8.4. Bust of Aristotle. Marble, Roman copy after a Greek bronze original by Lysippos, 330 BC. Roman National Museum of Altemps Palace, Rome.

Prior to the 20th century, it was feasible to believe that all celestial bodies were part of the Milky Way Galaxy. Although this belief existed into the 20th century, tools for measuring the great distances of cosmic bodies beyond our galaxy were under development. New discoveries opened the known universe to astonishing new interpretations. In 1925, Edwin Hubble, speaking at the Oxford Halley Lectures, unveiled the idea of other galaxies outside our own. From findings Hubble later made in 1928 (but not published until 1929), the details of Hubble's work, along with that of Harlow Shapley in locating the center of the Milky Way Galaxy a decade earlier, began the modern concept of the universe.

Throughout history theologians, philosophers, and astronomers have puzzled over the relationship of humanity to the cosmos, seeking to resolve known physical existence with that of the greater unknown universe. Many of the greatest thinkers have influenced this quest for understanding. Four in particular, two from the distant past, Ptolemy and Copernicus, and two of our own time, Shapley and Hubble, had significant impacts on the understanding of the universe. The consequences of their work form the history of how we have perceived ourselves in the cosmos.

77

Claudius Ptolemaeus

Known as Ptolemy, this astronomer, mathematician, geographer and visionary flourished in 2nd century Greek Alexandria when it was at the center of world culture and learning. We know little about Ptolemy personally. Yet his surviving writings include an eight-volume work on geography, three books on music, several on geometry, and at least five books on optical phenomena, including the first systematic study of the refraction of light. His *Almagest* encyclopedia established the Ptolemaic system that placed Earth at the center of the universe, where it remained at least until the 1500s. Ptolemy believed that the stars existed fixed in layers of crystalline spheres, each sphere outside the other, all revolving around Earth, and kept in motion by the Prime Mover of the universe. Although hopelessly incorrect in his geocentric interpretation of the structure of the universe, Ptolemy had the mathematical ability and scholarly reputation — backed up by the writings of the great Aristotle — to convince the world for more than 13 centuries that the universe existed as he envisioned it.

78

Nicolaus Copernicus

Copernicus ended the Ptolemaic geocentric model of astronomy by showing that Earth was not at the center of the universe, but was instead part of a heliocentric solar system with the Sun at the center.

Of Polish descent, Copernicus studied mathematics, law, and medicine. He devoted much of his adult life as the attending physician to a wealthy uncle who raised him and provided for his education.

Ironically, many of Copernicus' ideas about the solar system and the universe came from his reading of the classical Greek philosophers who held views different from those of Aristotle. Copernicus finished his groundbreaking book titled *De Revolutionibus* [On the revolutions of the Celestial Spheres] in 1530 at the age of 57 (**8.5**). Because his arguments favoring a heliocentric universe were in opposition to the teachings of the Roman Catholic Church, his book lay unpublished for thirteen years. He allegedly received his copy of the published work on May 24, 1543, the day he died. The uncle to whom he was so devoted was himself a Roman Catholic Bishop, and many believe that Copernicus delayed publication of *De Revolutionibus* intentionally to avoid persecution by the Church. In the book's preface — written to Pope Paul III — Copernicus made a veiled plea for the understanding of the pope, asking the pope to bear patience with what many expected the Church to view as a heretical idea. At one time, a fake preface even appeared inserted to replace Copernicus' original one. The newly inserted preface rationalized the book as merely a hypothesis, put forward for the sake of easing astronomical calculations. The motion of the moving Earth, it said, therefore did not require serious consideration as something that might actually be true. Such deception of the Holy See apparently spared Copernicus and his circle of followers from the indictments of religions infidelity that befell Galileo Galilei a hundred years later. It was the theory of

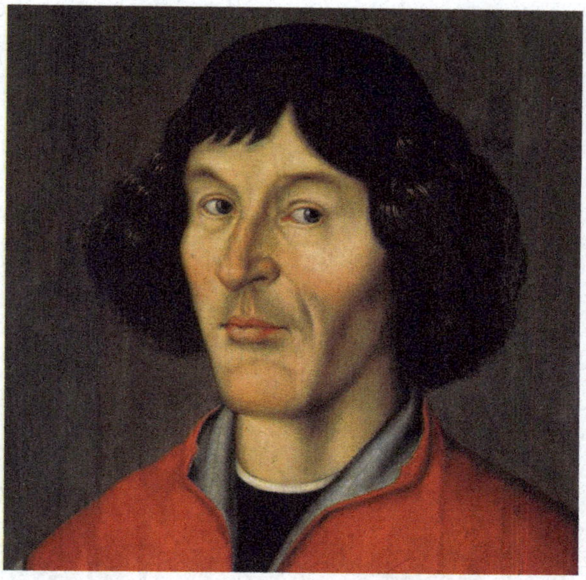

8.5. Portrait of Nicolaus Copernicus. The "Torun portrait" by an anonymous artist, c. 1580), Toruń Regional Museum, Poland.

Copernicus that greatly influenced the astronomy of Johannes Kepler and Galileo. Kepler first formulated the laws of planetary motion around the Sun while Galileo opened broad expanses of the universe through the development of the telescope. Without the pivotal work of Copernicus, however, it is doubtful that subsequent exploration of the universe would have produced the same results in the same historical period.

8.6. Portrait of Harlow Shapley. 1914. Mount Wilson Observatory.

79

Born in a small Midwestern town, Princeton educated scholar Harlow Shapley became one of the foremost American astronomers of the 20th century (**8.6**). Shapley was one-half of the 1920 Hale Lectures Great Debate on the scale of the universe, the other being the great Heber D. Curtis. Although no one now living attended that debate, and the public took little notice of it at the time, the debate stands today remembered as the point at which Shapley got us out of the center of the universe.

Curtis argued that the Sun was near the center of our galaxy and that there were other galaxies. Immanuel Kant, the noted 18th century German philosopher, had advanced the idea of such "island universes" in 1755. Shapley, on the other hand, countered that the solar system was not at the center of the Milky Way Galaxy but more toward its edge. He believed, however, that the galaxy was synonymous with the universe, and that other bodies visible in the universe were actually within our own galactic space. Both men, it would turn out, were right and wrong at the same time, as subsequent discoveries proved.

80

Edwin Hubble

Meanwhile, the pioneering work of Edwin Hubble in the measurement of distances and velocities of the galaxies opened the way to a universe much larger than either Shapley or Curtis had ever imagined. When Shapley received word from Hubble that he, Hubble, had proved the existence of galaxies outside the Milky Way, Shapley allegedly remarked to a close friend that Hubble's letter had destroyed his universe.

8.7. Portrait of Edwin P. Hubble. Photograph by Johan Hagemeyer, 1931. The Huntington.

Hubble, a graduate of the University of Chicago, was, like Shapley, born also in a small rural Midwest town. Hubble interrupted his early career in astronomy to serve in World War I, rising to the rank of major as an infantry commander, while winning the Purple Heart for wounds received in the French campaign (**8.7**).

Both Hubble and Shapley worked together for a short time at Mount Wilson Observatory. Because of incidents arising at Mount Wilson while Hubble was away serving in the armed forces, the two were never on amicable terms afterwards. Therefore, Hubble must have taken special delight in shattering Shapley's universe, while giving credence to the arguments of Shapley's rival, Curtis, who had argued against Shapley in the Great Debate for the existence of other galaxies like our own.

Hubble, Shapley, and Curtis were each a product of the Midwest, and all were born within 21 years of each other. Two of them, Hubble and Shapley, were from small rural towns in the same state of Missouri. These two men, who would do much to unveil the vastness of the universe, were born just four years apart and within less than 100 miles of each other.

Edvard Munch. *Starry Night.* 1893.
Oil on canvas, 53 ½ × 55 ¼ in. (135.9 × 140.3 cm.)
The Getty Art Museum.

X

INSIDE THE GALAXY

81

The Milky Way is that part of the Milky Way Galaxy that is visible from Earth. We see the stars, dust, and gas visible from our position inside the galaxy. Figure **9.1** is a view of the Milky Way. The center of the galaxy is toward the center of this composite photograph. In cross section, the Milky Way

9.1. The Milky Way. The constellation Sagittarius marks the approximate center of the galaxy in an image made up of a composite of photographs to create a wide-angle panoramic view. Photo by Nick Risinger.

appears as a ragged, irregular band of light seen from Earth stretching across the sky from one horizon to the other, beginning and ending at points 180 degrees apart.

The angle of orientation of Earth to the ecliptic — the plane of Earth's path around the Sun — and the angle of the ecliptic to the galactic plane, introduce an eccentric appearance of movement to the Milky Way at different times of the year. It appears sometimes perpendicular to the ecliptic, other times approaching, but never quite reaching, a parallel position to the orbital path of the planets.

Keeping in mind that the Milky Way is actually a cross-section through the disk of the Milky Way Galaxy; the gyroscopic motion from our point of view within the galaxy renders the scene of the Milky Way like a slow motion carnival ride on a tilt-a-whirl. In **9.2,** wide-angle photographs show the location of the Milky Way in the

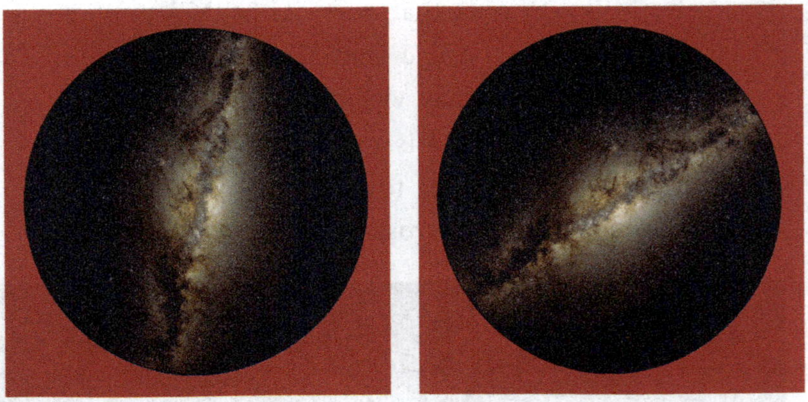

9.2. The movement of the Milky Way.

month of May and how much it has tilted a month later in June.

Looking out from Earth, the plane of the Milky Way Galaxy cuts across the ecliptic path of the solar system at approximately a 60-degree angle, which means that the solar system sets cockeyed in the galaxy, tilted on its edge in the galactic disk. At the same time, the axis of Earth, whose poles and equator define the celestial sphere, tilts 23.5 degrees from the ecliptic path of the solar system. This tilt, which is also responsible for Earth's seasons, causes the band of the Milky Way to appear at one angle and then another over the course of a year's time.

82

The galaxy

The Milky Way is visual proof that we are inside the Milky Way Galaxy. From inside the galaxy, modern astronomers are able to estimate the size of the galaxy and the orientation of Earth and the solar system within it. However, where we are located in the galaxy sets limits of our point of view, not only of the galaxy but also of the universe itself. The irregular pattern of the Milky Way, which is its spottiness, is due to interstellar dust that obscures the view along the plane of Earth inside the galaxy. Looking perpendicular to the galactic disk, that is, away from the plane of the galaxy's disk, we can see great cosmic distances. However, our view along the plane of the disk, because we are inside it, is limited to less than one-tenth the distance to the center of the galaxy. From our vantage point inside the disk, we see less than 1.0 percent of the existing galaxy. Only a narrow arc of a single arm of the Milky Way Galaxy is what we see as the Milky Way.

83

The center of the Milky Way Galaxy is toward the constellation Sagittarius, located in the southern sky along the ecliptic path. Sagittarius appears clearly during midsummer evenings. When we look toward Sagittarius, we are looking into the center of the galaxy.

9.3. Infrared image of the Milky Way Galaxy. Infrared photography cuts through the dark dust lanes and clouds to reveal the galactic plane and the bulge of its core. Photo by J. M. Carpenter and R. Hurt.

Although we cannot begin to see the center of the galaxy, scientists can detect the effects of its intensity. The center of the galaxy is radiating invisible energy at non-optical levels. The remarkable image in **9.3** exposes the center of the Milky Way Galaxy. Invisible radio, microwave, and infrared energy make it possible to *see* into the center of our galaxy, something that is impossible with light energy, because interstellar matter blocks the optical view of the galaxy center. The galactic disk flows outward from the bright galaxy nucleus as if in cross-section.

The interstellar matter between Earth and the rest of the galaxy that limits the range of sight into the galactic disk creates a cone of visual extinction from Earth's position inside the galaxy. This cone of extinction allowed American astronomer Harlow Shapley to make his famous discovery of the center of the galaxy.

Shapley's discovery, a result of visual observation and brilliant scientific deduction, changed forever the perception of our place in the galaxy. In the process of plotting the locations of clusters of stars on a map of the universe, Shapley recognized that they formed a semicircular pattern that we now know defines the galaxy center. Star clusters — called globular clusters — belong to the galaxy and are among the oldest parts of it. Clusters lie in a spherical pattern about the galaxy's radius. Because dust and debris visually black out the center of the galaxy, Shapley's brilliance was in recognizing that the spherical pattern of star clusters outlined the nuclear bulge of the center of the galaxy. We observe no similar pattern above or below the plane of the galaxy where the view of the universe is virtually unobstructed.

Huge portions of the universe are unseen because of the cone of extinction. Vision is restricted in the direction of the nuclear bulge at the axis of the galaxy, a bulge beyond which we cannot see. We will not have a good view of what is out beyond the other side of the galaxy for another 100 million years when the solar system revolves in the galaxy to a point opposite its present location.

84

Racing through the galaxy

The Sun and the solar system revolve around the central axis of the galaxy much as the planets orbit the Sun. The speed at which the solar system orbits the galactic center is more than 200 kilometers per second. When taken out of context this is a meaningless number.

Object	Km/hr.	Mi/hr.	Km/min.	Mi/min.	Km/sec.	Mi/sec.
Galaxy through the universe	2,160,000	1,340,000	36,000	22,360	600.0	373.0
Sun around the galaxy	720,000	447,000	12,000	7,452	200.0	124.0
Earth around the sun	108,000	67,000	1,800	1,118	30.0	18.6
Galileo Jupiter spacecraft	80,000	50,000	1,300	830	22.2	13.8
Satellite orbit around earth	29,000	18,000	480	300	8.1	5.0
X-43 Hypersonic aircraft	11,760	7,307	196	122	3.3	2.3
Moon around Earth	3,700	2,300	62	38	1.0	0.6
Bullet from small caliber rifle	3,240	2,012	54	34	0.9	0.6
Earth's rotation	1,670	1,035	28	17	0.5	0.3
Sound velocity	1,225	760	20	13	0.3	0.2
Airliner cruise speed	900	560	15	9	0.25	0.16
Race car	350	215	6	4	0.10	0.06

Table 9.4. Approximate velocities of objects in kilometers and miles.

Nothing in our experience on Earth travels at anything approaching a velocity of 200 kilometers per second. It would be like flying from Washington, D.C., to Philadelphia in one second. To underscore this point, a few velocities appear in Table **9.4**. A bullet from a small-caliber rifle, for example, travels at 0.9 kilometers per second, about the speed at which the Moon orbits Earth (**9.5**).

Commercial airliners cruise on average at about 600 kilometers per hour, about 1/6 of a kilometer per second. A racecar can travel at 350 kilometers per hour, but in seconds that is equal to 1/10 of a kilometer per second. At a speed of 350 kilometers per hour, a car can travel the length of a football field in one second; fast, but hardly competitive in a race through the galaxy. In the same amount of time, one second, a rider on Earth travels with Earth through the galaxy 200 kilometers.

9.5. High speed, Schlieren and Photo instrumentation photograph. Photo by Harold Edgarton.

Earth is in motion in many directions at once. Measured at the equator, Earth is rotating at 1,670 kilometers per hour, or about one-half kilometer per second. At the same time, it is orbiting the Sun at 30 kilometers per second and revolving around the galactic center at 200 kilometers per second.

Simultaneous movement of Earth is at three different speeds: (*a*) the revolution of Earth at 1,670 kilometers per hour; (*b*) the orbital speed of the planet around the Sun at 108,000 kilometers per hour; and (*c*) the velocity of the solar system on its path through the galaxy 720,000 kilometers per hour. These movements are exclusive of the movement of the galaxy that is also speeding through the universe at even greater speeds.

If a person is in movement on Earth, a fifth relative velocity occurs, that of our own locomotion. We may add also the motion of Earth's precession through the ecliptic plane, as well as its eccentric movements around its Barycenter.

Nothing we know on Earth travels at the high speeds associated with our passage through the universe. Fortunately, from our position on Earth, we are not aware of the great speeds at which we are moving relative to gravity centers beyond our own. We feel our balance and sense of motion only relative to our motion on Earth.

85

The velocity of Earth through space measured against terrestrial air and ground speeds is remarkable.

The speed of sound at room temperature is 0.34 kilometers per second, slower than the rotation velocity of Earth. Earth is rotating faster than the speed of sound.

Hypersonic speed is Mach 5, or five times the speed of sound. The old X-15 rocket plane flew faster than Mach 5 at 2.1 kilometers per second, much faster than the speed of sound. At one time, NASA had in its plans an advance design airship, the X-34 that would fly 8.6 kilometers per second. That speed is nearly nine times the muzzle velocity of a small caliber rifle that has a speed of about 900 meters (0.9 kilometers) per second.

9.6. NASA X-43 unmanned hypersonic aircraft recognized as the fastest aircraft ever made. NASA.

Designers of the proposed X-34 hypersonic shuttlecraft envisioned it would someday travel at a velocity almost ten times the speed of a bullet. The X-34 would fly into outer space and return, thus replacing the decades old space shuttle in a major advancement of space travel.

In 2001, NASA announced its termination of the X-34 project, saying only that technology had not yet advanced to the point of realizing such an advanced design. After just five years of development, planners concluded that such futuristic travel would remain left as a challenge to future generations and opted for less complicated space technology. Nevertheless, new initiatives in hypersonic flight continued. The unmanned X-43 model test flew at Mach 9.6, the fastest aircraft ever flown (**9.6**). The X-43 series also ended with adoption by NASA of new technology to continue high speed unmanned flight in a different experimental series.

In 1969, it took the Apollo 10 spacecraft a little more than three days to reach the Moon traveling at an average speed of about 5,000 kilometers per hour, or 1/6 the projected velocity of the abandoned X-34. The X-34 would conceivably have gone to the Moon and back in a day. Still, for all its technological advancement, the X-34 would have traveled at a velocity less than half the speed of Earth around the Sun.

Beyond the gravitational field of Earth, manmade high velocity travel is already a reality. Impressive cruise speeds recorded by such spacecraft as the Galileo Jupiter mission probe exceeded velocities of 80,000 kilometers per hour. Very soon, in-space technology will enable construction of advanced space machines using inertia engines whose design will not have to accommodate the propulsion requirements necessary to break Earth's gravity. Just how such ships will land and take off from other Moons and planets, however, is yet another problem.

The incredible array of velocities as measured relative to Earth serves to emphasize that Earth is moving through space at a rate almost beyond comprehension, and well beyond anything related to the common individual's experience with movement and speed. In the solar system, we are traveling at a velocity twice as fast as any projectile in human experience or envisioned for human experience, and in the plane of the galaxy, the velocity of Earth is 25 times faster than any moving object known on Earth, literally 200 times faster than a bullet. Add to all this that the galaxy is itself hurtling through the universe at an incredible speed. One begins to appreciate fully the function of the inner ear that knows no motion except in relation to Earth's gravitational field.

86

Even at the Sun's amazing velocity of 200 kilometers per second, it still requires at least 200 million Earth years for the solar system to complete one revolution around the galaxy. If estimates of the age of the Sun are correct, that is 4.4 to 5.5 billion years, the Sun has made just 25 trips around the galaxy since its beginning. Thus, the Sun is 25 galactic years old, give or take three to five galactic years.

The last time the Sun was in the position it is in now was during the Mesozoic era, the third era of geologic time, a time characterized by its reptilian life. The dinosaurs came into existence on Earth and disappeared while the Sun made its last revolution around the galaxy. Energy now arriving from the galaxy's center began its journey 30,000 years ago. During this time, the solar system moved 200 trillion kilometers in its orbit around the galaxy. In the same amount of time, the Sun traveled 5,000 times farther than any of the planets in orbit around it, yet the distance it traveled in the galaxy was only 1/5,000 of the solar system's orbit around the galaxy.

In the 30,000 years required for energy to travel from the center of the galaxy to the solar system and Earth, the entire history of humankind took place. This minuscule segment of galactic time cannot be represented graphically even by the thickness of a line pointing to galactic year zero on the galactic clock. The existence of humanity has progressed from zero less than the thickness of a line. We will not reach galactic hour 100 million on the other side of the galaxy, until Earth year 99970000. Meanwhile, before the appearance of humans,

Earth already had made more than 24 trips through the galaxy without us, and the trip we are on now barely has begun.

87

Beacons of time

The real time experience of viewing events spread across millennia is one of the great fascinations of astronomy. Light is like a time machine. It permits the observation of events in space that occurred in the distant past. A star ten light years away conveys through its light a message of what it looked like ten years ago. Light has a finite speed. Because of this, it requires time to reach its destination. Any object at a great distance will experience a time delay. For example, we view the Moon on Earth, as it appeared 1.3 seconds ago; the Sun 8.3 minutes in the past; the planet Pluto about 5.5 hours ago; and the nearest star Alpha Centauri 4.3 years ago. The star Sirius appears to us as it looked 8.6 years in the past; the star Aludra 2,500 years ago, the galaxy Andromeda 2.2 million years ago; and on and on, to the farthest reaches of the universe.

Light reaching Earth from a cosmic body may be associated with an event on Earth that happened at approximately the time the light's journey began. The Andromeda Galaxy, for example, as seen today was the Andromeda galaxy as it existed at approximately the time of the emergence of the earliest bipedal hominids on Earth more than two million years ago.

Each star we see is a different beacon of time. No two are of the same visual age. For instance, in the northern hemisphere, looking southeast in the winter sky, is the constellation Canis Major, the Big Dog (**9.7**). In it is the star Aludra. Visible in the winter Milky Way low in the southern sky, Aludra lies just south of the star Sirius the brightest star in the sky. Only the seventh brightest star in Canis Major, Aludra appears today, as it actually existed 2,500 years ago, a mere blink in the Sun's

revolution around the galaxy. The light from Aludra just reaching Earth today began its journey at the time the Etruscans flourished, before the founding of Rome, and long before the Roman Empire rose and fell. The light that we see began concurrently with the birth of Buddhism on one continent and the flowering of the Olmec culture on another, starting fully five centuries before the birth of Jesus and more than two and a half centuries before Alexander spread the Aristotelian worldview throughout the Greek empire. The light of Aludra that is reaching Earth now has traveled through much of the history of the civilized world. After a journey of nearly two and one-half millennia, light that left Aludra in the 6th century B.C.E. finally reached the star Sirius, and then continued the remainder of the trip from Sirius to Earth in a comparatively minuscule 8.5 years. These two stars, Aludra and Sirius, though they appear relatively closely spaced in the celestial sphere, exist greatly removed from each other in the vast depth of the universe. Like the Big Dipper, the radial locations of stars in any part of the sky are seldom what they appear to be.

9.7. The stars of constellations. Aludra in Canis Major is at bottom left The map is looking south at approximately 9 p.m. After Bob King.

88

We are travelers together in a distant place, moving through space and time relative only to points known to us. The stars and the galaxies are our connections to the cosmic past. The farther out they are, the farther back it is possible to see into the history of the universe. Much of the history of the universe remains unknown to us, beyond a time boundary where we cannot look, where cosmic bodies lie at such great distances that their light travels through space not yet detected on Earth.

Scientists on another planet, in another galaxy billions of light years away, one day may see our galaxy. They will analyze its composition and wonder about the possibilities of life like theirs deep in the outer reaches of space. They will do this millions of years after our Sun has disappeared. Perhaps they will see the light that left the galaxy in Aristotle's time, or when the dinosaurs roamed Earth. Light tells only of the nature of its source at the time it left and nothing about the light's place of origination in the thousands, millions, or even billions of years since it began its journey through the universe.

THE ILLUMINATED UNIVERSE

We have a hunger of the mind which asks for knowledge of all around us, and the more we gain, the more is our desire; the more we see, the more we are capable of seeing.

Maria Mitchell.
Astronomer, 1818-1889.

Jean Edme. *L'Astronome* Nochez. 1750–1800. (After François Cotiber.)
Etching, $9\ ^{13}/_{16} \times 7\ ^{1}/_{16}$ in. (24.9 × 17.9 cm.)
The Metropolitan Museum of Art.

XI

MEASUREMENT OF THE STARS

89

Throughout history, people have been fascinated with stellar distance; how far away are the stars? One person who returned to this subject repeatedly in his work was the artist Vincent Van Gogh. Like many artists, Van Gogh was enchanted with the evening sky, and on more than one occasion made the stars the subject of his paintings. His painting *Starry*

10.1. Vincent Van Gogh. *The Starry Night*. 1889. Oil on canvas, 29 x 36 ¼ in. Museum of Modern Art.

Night is among the most celebrated paintings in the history of art (**10.1**).

Writing to his brother Theo, Vincent, looking up at the stars, reflected, "Why, I ask myself, shouldn't the

10.2. The star Alpha Centauri, the brightest star in the constellation of the Centaur and our nearest star neighbor. It is the same type of star as the Sun, and visible only from southern latitudes south of 25 degrees. European Southern Observatory.

shining dots of the sky be as accessible as the black dots on the map of France? Just as we take the train to get to Tarascon or Rouen, we take death to reach a star." In their "exaggerated brilliance," Van Gogh imagined each star represented "eternity in its place above this world."

The nearest star to our solar system is Alpha Centauri, the brightest star in the constellation of the Centaur (**10.2**). Alpha Centauri is 4.3 light-years away, or about 40 trillion kilometers. To give some idea of how far that is, it would take 270,000 trips to the Sun to equal just one trip to Alpha Centauri. Traveling at 40,000 kilometers per hour, or 33 times the speed of sound, it would require more years to reach Alpha Centauri than has been spent in the entire development of the modern human species, spanning more than 100,000 years. Moreover, this is our *nearest* stellar neighbor in the universe.

90

A star is a self-luminous mass of hot gas held together by its own gravity. The universe contains trillions and trillions of these celestial bodies. The Milky Way Galaxy alone contains more than 100 billion stars. All the stars that we see when we gaze into the sky with the unaided eye are in the Milky Way Galaxy. We cannot see individual stars in other galaxies because they are too far away.

It is not only impossible to see stars outside our galaxy; it is possible to see only a limited number of them within the galaxy. Most of the stars in the Milky Way Galaxy are either too far away or too faint to see. Of the billions of stars in the galaxy, relatively few of them are within the visual field. Only about 6,000 stars are visible to the unaided eye under the best of viewing conditions.

When one looks into a clear evening sky, the impression is that there are more than 6,000 stars in our visual field. However, there are not. In fact, from a vantage point in any hemisphere, depending on one's location, fewer than 3,000 individual stars may be visible. In addition, if one considers our limited angular view, that is, we cannot see behind us, the number of stars in the visual field at any one time is probably no more than around 1,000. The entire star catalog of Hipparchus, the

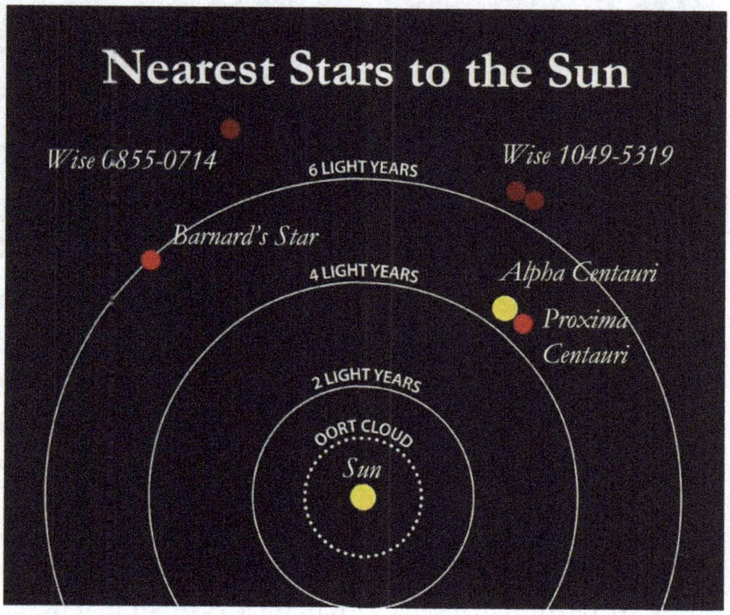

10.3. Nearest Stars to the Sun.

Greek astronomer and careful observer of the heavens, listed only 850 stars visible to the human eye. Fewer than 50 of the brightest ones attract our attention.

Therefore, depending on what part of the sky is in view, we may be aware only of a few stars out of the 3,000 or so that may be visible to the unaided eye. Only a few of the brightest ones draw our interest.

91

Two things occupied much of the early science of cosmic theory: the motions of cosmic objects and the distances to them. Astronomy has not succeeded fully yet in measuring either motion or distance.

10.4. Giovanni di Paolo. *Expulsion of Adam and Eve from Paradise.* 1445. Tempera and gold on wood. 18 ¼ x 20 ½ in. (46.4 x 52.1 cm.) The Metropolitan Museum of Art.

Many of the mistakes of the early astronomers in measuring stellar distances stemmed from the application of terrestrial distance measurement techniques to the measurement of the stars. What worked on Earth would expectedly hold true beyond Earth, they reasoned. After all, as late as the 18th century, astronomers still assumed that the stars were relatively close to Earth, encased in their own revolving spheres, each spherical layer being progressively more distant from the central Earth. Even the great Kepler allegedly believed that all stars existed in a shell a few miles thick. The artist Giovanni di Paolo celebrated this ancient idea in his 1445 painting *Expulsion of Adam and Eve from Paradise* in (**10.4**). Di Paolo's painting followed the Aristotelian and medieval concept of the universe, whereby Earth was at the center of a series of concentric spheres. In Giovanni's painting, Aristotle's original idea of multiple cosmic layers received a Biblical setting; God

appears depicted banishing Adam and Eve from his creation. The same theme appeared in a 13th century manuscript illumination in which God appeared just beyond the stars (**10.5**). Today, the understanding of the placement of the stars is radically different from Giovanni di Paolo's 15th century fantasy, and the religious constraints of 13th century monks. The stars are not close to Earth at all.

10.5. God measuring the universe at the instant of creation. Codex 2554, (attributed to Reims, France). 13th century. National Library, Vienna.

Knowledge of stellar distances is comparatively recent. The distances to the Sun and planets have been known for centuries, but the first successful attempts to measure the distance to stars beyond the Sun occurred only in 1838. Using a simple method of measurement called *parallax*, the first rough discoveries of the vastness of the universe entered the cosmic equation. Since then, with each new estimate of stellar and galactic distances, the size of the universe grew.

92

Light-years

The closest star to Earth is the Sun. Its mean distance is 149,597,870 kilometers. Beyond the Sun, the nearest stars are so far away that kilometers become insignificant as a tool of measurement. In astronomy, time and distance uniquely combine in a single measurement, the light-year. The light-year expresses the size of the universe beyond the comparatively small confines of our solar system.

The distance to a star measured in light-years is the number of earth-years that light will need to travel from that star to reach Earth, traveling at a velocity of 300,000 kilometers per second, or the speed of light. Since one light-year is equal to approximately 9.46 trillion kilometers, a star that is ten light-years away, for example, is at a distance of 94.5 trillion kilometers from Earth. Such a distance is unimaginable on a terrestrial scale. Were the journey to such a star possible, it would take more than 13 million years by modern jet travel to go a distance of 10 light-years.

Most stars are beyond 10 light-years. There are fewer than 30 stars within 15 light-years; only 18 are within 12 light-years, and a mere seven stars are inside 10 light-years away. Some of the distances to stars are so great that scientists believe there are stars seen now as points of light in the nighttime sky that no longer exist; only their light, which began a journey through space millions of years ago, remains to convey the messages of their past realities. Powerful telescopes are able to detect light from galaxies that are so far away that their light seen today may actually be that associated with events beyond the time of the beginning of the solar system and approaching the birth of the universe.

Using sophisticated new measurement techniques, observers predict that the farthest detected cosmic bodies are approximately 14 billion light-years away. Light seen coming from these objects began a journey 14 billion years ago, well before the formation of the solar system and long before the beginning of Earth. Hundreds of similar objects appear in the image in **10.6**. Each point of light in this photograph is its own source of light and energy.

10.6. View of the evening sky. This image of a tiny portion of the sky is typical of the array of stars seen on a clear night. NASA. Photo by Jim Misti.

Where the sky appears black and empty to the unaided eye, telescopes reveal layer upon layer of points of light. Hardly any of the objects seen in the photograph in **10.6** are stars, and none is a planet. Planets that exist outside our solar system are much too far away to be visible. Instead, each point of light in this tiny sample of the universe represents billions of stars in galaxies and clusters of galaxies as far into the universe as it is possible to look.

93

The hemisphere of Eratosthenes

Among the earliest techniques for measuring great distances on Earth, one incorporated a method using the shadow rod. The shadow rod was an ancient measuring device dating at least to the Babylonians and Egyptians. They used it in land surveying and measurement. The shadow rod technique served as a precursor to what later became the parallax method of distance measurement.

The parallax method eventually would require the invention of trigonometry, a mathematical application, whereby enough known sides and angles of a triangle will result in calculation of the remaining sides and angles. Trigonometry in the Greek means "angle." However, before the invention of trigonometry, the ancients did not have the benefit of such a shortcut method of calculation and had to rely on direct methods of geometric measurement. Descriptions of two such ingenious methods used more than two thousand years ago are worth recounting.

94

In the third century B.C.E., the Egyptian-Greek Eratosthenes, Librarian of Alexandria, used the shadow rod to measure the circumference of Earth. His writings are lost, so the details of his method are unknown, but it is certain that he used some type of shadow rod technique involving the triangular measurement of two shadows and the distance between them.

As early as 2000 B.C.E. the Egyptians had arrived at measurements for the circumference of Earth. Later, Aristotle and Archimedes, the latter a contemporary of Eratosthenes, each published measurements, although their separate findings appear to be of dubious origin and possibly borrowed from the Egyptians. The measurements of Eratosthenes are the most accurate.

Eratosthenes understood Earth to be round. He knew therefore that the measurement of a segment of

Earth's circumference presaged a way to calculate its full circumference. He observed that at summer solstice when the Sun was directly overhead at Syene, Egypt (now Aswan), it cast no shadow. At the same time, 800 kilometers due north at Alexandria, a shadow was measured to be a little more than 7 degrees. At Alexandria, an accurate angle could be determined by simply leaning the shadow rod into the Sun until it cast no shadow, then measuring the angle at which the rod was out of plumb. This may or may not have been Eratosthenes' method. According to one commentary, he mounted the shadow rod on top of a hemisphere to represent one-half the circumference of Earth. The length of the shadow cast on the hemisphere was proportional to its relationship to Earth's surface. Thus, Eratosthenes would not have needed to employ angular distances at all. Eratosthenes realized that shadows from the Sun play over only one-half of Earth at any one time. Rather than applying his results to a sphere, he used a hemisphere which when doubled obviously would be proportional to the sphere of Earth. It is doubtful that he employed degrees at all but instead applied the rules of proportionality common to the scholars of his day.

Eratosthenes had the benefit of Euclid's great work on geometry, *The Elements*. He had knowledge on how to employ geometric calculation. He probably did not use the 360-degree circle. The Babylonians had developed such a circle from their base-60 numbering system and applied it to the circle of the Zodiac. However, the degreed circle was not a mathematical construct until a couple of generations after Eratosthenes. Nevertheless, Eratosthenes was able to deduce that his shadow rod cast a shadow 1/25 of the arc of a semicircle, or 1/50 of a circle. The length or angle of the shadow would thus represent 1/50 of the full circumference of Earth.

Eratosthenes reasoned, therefore, that if the shadow displacement from Syene to Alexandria was 1/25 of a semicircle, or 1/50 of a circle, then the distance segment from Syene to Alexandria was 1/50 the distance around Earth. Therefore, the circumference of Earth would be 50 times the distance between the two cities, which are 800 kilometers apart: (50 x 800), or 40,000 kilometers. Some believe that Eratosthenes' actual estimate was 40,500 kilometers. In any event, he was remarkably close to the known circumference of Earth today of 40,820 kilometers. Allowing for the fact that Eratosthenes recorded his results in stadia, the prevailing measurement of his day, and the fact that measurement standards and equivalencies have changed over the years, Eratosthenes' results are astounding.

95

The metric system did not originate until 1790 built upon the standardized kilometer determined to be 1/10,000 of the distance between the equator and the North Pole. Under the metric system, the polar circumference of Earth was, by definition, 40,000 kilometers; the length of a single quadrant was 10,000 kilometers. In recent years, recalibrated metric length uses a standard developed relative to the value of the speed of light. However, the new calibration did not significantly change the kilometer from its initial standard tied to Earth's quadrant.

Meanwhile, Eratosthenes recorded his measurement in the stade, the ancient Greek measure of distance. The stade, or stadium, derived from a racecourse, usually semicircular with tiers of seats for spectators. The stade was the set length of the racecourse thought to be about 185 meters. The correlation between stadia and the metric system is inexact. Some archeologists have suggested that, like the kilometer, the stade underwent adjustment from time to time and may even have had different values at different locations. Depending on what value of the stade is used, Eratosthenes' range of accuracy in predicting the circumference of Earth is anywhere from 0.5 to 16 percent.

Historians believe that Christopher Columbus had access to Eratosthenes' results, but made a mistake in converting stadia to the units of measurement of 15th century Europe. Columbus planned his famous 1492 journey around the world on a global circumference of 17,000 miles. Thus, upon arrival in America, he thought he had sailed much farther around the world than he actually had. The Indians he encountered were not of the East Indies, as expected, but of the West Indies of the New World.

Eratosthenes had many interests. He is credited with the measurement of the tilt of Earth's axis to a high degree of accuracy, and he even called for a leap day every fourth year to correct for the solar year. His accomplishments in Earth science earned for him the modern title of the founder of scientific geography. His map of the ancient world was the first to show lines of latitude and longitude. In addition, his method of identifying prime numbers we consider a benchmark of original thinking. According to the surviving commentaries about him, he wrote extensively on astronomy and mathematics, geography, philosophy, and history, and was a poet of some note. So varied were his interests — none of which he seems to have claimed as a specialty — that he came to be known as second best in everything and best at nothing, leading to the nicknames *All-rounder* and *Beta*, the second letter of the Greek alphabet.

Eratosthenes was born in Cyrene (now Libya) and educated in Athens. No one knows for sure just when he was born. Historians date his birth anywhere from 284 to 235 B.C.E. He died in Alexandria sometime between

201 and 192 B.C.E., probably around the age of 80. He allegedly starved himself to death in a period of despondency apparently over old age and blindness. Abundant commentaries have survived about his work and about him as a person to identify him as one of the great mathematicians and astronomers of all time.

96

The lunar dichotomy of Aristarchus

An early attempt to measure great distances beyond Earth also involved the use of shadows, but in an application that differed from the shadow rod technique of Eratosthenes. Aristarchus of Samos, the Greek astronomer and mathematician, in the 3rd century B.C.E. was the first to predict in astronomical terms distances to challenge the classically held notions about the scale of the cosmos. He was the first to describe the true heliocentric nature of the solar system. Aristarchus estimated the size and distance to the Moon; the size and distance to the Sun; and deduced correctly that the stars were like our Sun at great distances, not dots on a celestial sphere. His clear logic still stands as a model of brilliant deductive reasoning.

After careful observation of a lunar eclipse, when Earth is between the Moon and Sun, Aristarchus realized that Earth's shadow cast by the Sun would have a conical shape (**10.7**). He understood, too, that he could duplicate a similar conical shape by creating a false eclipse with a

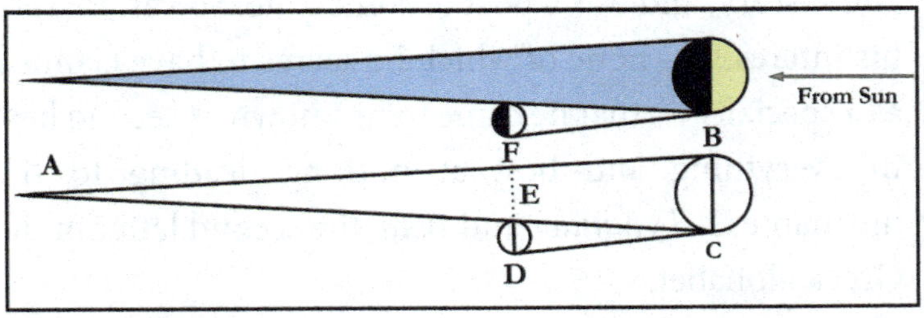

10.7. The lunar distance calculation of Aristarchus.

circular shape of a known size located at a specific distance from the eye until it just covered the solar disk, as would happen with the circular shape of Earth. As it turned out, a shape of one centimeter eclipsed the solar disk from view at about 108 centimeters. Using these same proportions, the point of Earth's shadow cone is 108 Earth diameters away. From this, Aristarchus could draw the eclipse diagram, with the point of the cone being the farthermost point at which Earth can block all the Sun's light.

The Greeks knew from Eratosthenes that the diameter of Earth was about 12,500 kilometers. Therefore, the length of the shadow cone, calculated in modern metric measurement, is 108 times 12,500 kilometers, or 1.35 million kilometers.

During a lunar eclipse, the Moon moves through Earth's shadow. Therefore, the Moon cannot be more than 1.35 million kilometers distant. Aristarchus estimated that it took about 2.5 diameters of the Moon to pass through Earth's shadow during a lunar eclipse. He knew also that the Moon just covered the Sun in a solar eclipse. He reasoned therefore that the shadow cone of the Moon was of similar shape as that of Earth and directly proportional to it. The problem then invited a solution using geometry. In figure **10.7**, the angle *ECD* is equal to the angle *EAF*. *FE* is the diameter of the cone of Earth's shadow at the distance of the Moon, or where the Moon passes through Earth's shadow. *ED* is the diameter of the Moon. We observe the ratio of *FE* to *ED* to be 2.5:1, or 2.5 times the diameter of the Moon. Based on similar isosceles triangles *FAE* and *CDE*, *AE* is 2.5 times the length of *EC*, from which it is deduced that *AC* is 3.5 times as long as *EC*. *AC* has an estimated dimension of 108 Earth diameters, or 1.35 million kilometers. Therefore, the distance to the Moon is 1,350,000/3.5 or about 385,000 kilometers. This estimate proved to be very close to the actual distance of 384,400 kilometers. Given that the visual estimate of the Moon's diameter passing through Earth's shadow was prone to error, Aristarchus' ancient calculation is as remarkable as Eratosthenes' measurement of Earth. Of course, once the geometry of the lunar eclipse was drawn, the diameter of the Moon followed.

Next, Aristarchus tackled the distance to the Sun. In this, he was less successful, but managed in the process to show that the Sun was much larger and at a much greater distance than previously thought. Again, shadow played a key role in his method.

Here he reasoned that when the Moon appears to be half-full, the line from the Moon to the Sun must be exactly perpendicular to the line from the Moon to the observer (**10.8**). By carefully measuring the angle between the half Moon and the Sun, he was able to construct a thin right triangle with the baseline being the newly determined distance from Earth to the Moon.

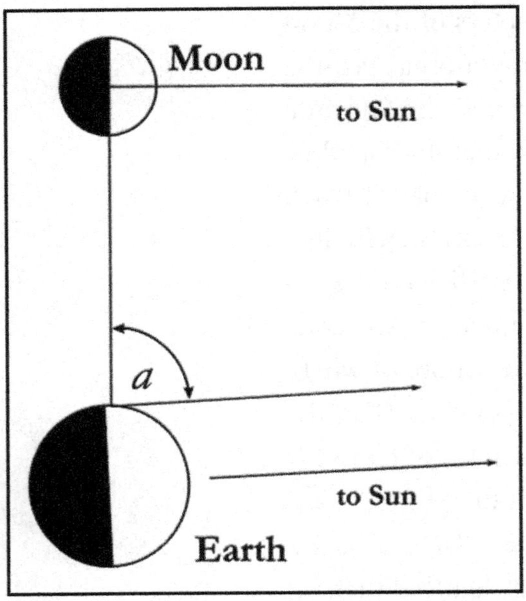

10.8. Aristarchus' distance to the Sun calculation.

Using geometric calculation, he estimated the distance to the Sun to be 20 times the distance to the Moon, and its size as 20 times the diameter of the Moon. The angle used in setting up this triangular proportion was off by quite a bit because it is actually 0 degrees 10 minutes and not the 3 degrees, or 1/30 of a quadrant he used. A quadrant is one-fourth of a 360-degree circle. In his system, 1/30 of a quadrant would be equivalent to 1/10 of a station of the Zodiac, which would have been Aristarchus reference. He would not have used degrees or trigonometry since neither was yet in use in his time. Because of his inaccurate angular measurement, Aristarchus' estimates regarding the Sun fell considerably short of the true distance and size relationships. The proportional distance and size of the Sun, relative to the distance and size of the Moon, is more on the order of 400 times and not 20 as he predicted. Although his measuring devices were inadequate, and his results were wrong, the procedures he followed were highly original. They led to what later would become the stellar parallax method of measuring the stars. Still, even based on his faulty estimates, Aristarchus was able to conclude that the Sun was substantially larger and at a much greater distance than anyone had previously envisioned. Moreover, because he discovered the Sun to be substantially larger than Earth, he knew that it could not orbit Earth as was believed at that time. Thus, Aristarchus proposed his heliocentric solar system.

Modern astronomers regard Aristarchus as one of the greatest thinkers in human history. Yet we know little about his life. Born approximately 310 B.C.E., he lived on the Greek island of Samos, located just off the west coast of present-day Turkey, at a considerable distance from both Athens and Alexandria, the political and intellectual centers of the Greek Empire. He was of the Pythagorean school of thought that employed geometrical and arithmetical relationships to explain the universe. He allegedly studied at Alexandria, probably before Eratosthenes' time, although their two lives overlapped some forty years. They no doubt knew each other's work and may even have collaborated at one time. Aristarchus came up with a more accurate calculation of the Eratosthenes leap year of 365 ¼ days.

We know Aristarchus' work mostly through the writings of Archimedes and Plutarch who refer to a book on heliocentric theory that has since been lost. The only work of his that survives is a treatise on *The Sizes and Distances of the Sun and Moon* in which he describes his remarkable geometric argument. Unfortunately, he makes no mention in this work of the Sun-centered theory referred to by Archimedes and Plutarch that later influenced the heliocentric theory of Copernicus.

The genius of Aristarchus is highly regarded today, credited with being the first to establish the correct relationships of Earth, Moon, Sun, and stars. Despite his acclaim in recent times, his peers roundly criticized him during his lifetime for his theories, which, like the theories of Nicholas Copernicus who revived his ideas 17 centuries later, were outside the mainstream of Greek philosophical thought and directly contradicted the Aristotelian doctrine of an Earth-centered cosmos. Archimedes, having reported the theories of Aristarchus in the *Sand-Reckoner*, criticized his views saying they conflicted with the more common teaching of the astronomers. Cleanthes, a contemporary of Aristarchus in Greece, went so far as to write a treatise "against Aristarchus," in which he charged Aristarchus with impiety for putting in motion the "Hearth of the Universe Earth," by supposing heaven to remain at rest and Earth to revolve. Archimedes and Cleanthes mimicked Dercyllides, among others, who expressed the general opinion of the time when he wrote,

"We must suppose Earth, the Hearth of the House of the Gods, according to Plato, to remain fixed, and the planets with the whole embracing heaven to move and reject the view of those [Aristarchus] who brought to rest the things which move and set in motion the things which by their nature and position are unmoved, such a supposition being contrary to the theories of mathematicians."

It is very doubtful if any actions ensued against Aristarchus for his opinions in third century Greece, as did occur 18 centuries later against Galileo for holding a similar view. Nevertheless, condemnation of Aristarchus was by both religious dogma and mathematical scrutiny for his teachings that Earth circles the Sun. Even the

Great Hipparchus, thought to be the greatest astronomer of antiquity, rejected the Aristarchus theory because he thought the angles of the stars would change if Earth actually moved. No one else understood then except perhaps Aristarchus that the stars are at such great distances as to produce no perceived angular changes.

Ironically, Aristarchus' detractors in stating their case against him wrote a clear description of the heliocentric system as it proved ultimately to exist. They chided Aristarchus for believing that the Sun was one of the fixed stars, the closest to Earth, and for supposing heaven to remain at rest and Earth to revolve in an oblique circle while it rotates at the same time about its own axis, causing day and night. They could have added that Aristarchus knew, too, that Earth's axis was inclined to the plane of the ecliptic, causing the change of seasons. His ideas, if not his measurements, were precisely accurate.

Some think Aristarchus may have purposely misrepresented his predictions about the scale of his heliocentric system, anticipating that his measurements would appear truly unbelievable to the people of his time. He seems to have been more pleased with the creative mathematics of his methods than the accuracy of his data. In that respect there is no question that his discoveries stand as pillars of brilliant deductive reasoning.

97

The decline of Heliopolis and Persepolis

Long before the Greeks, the Egyptians understood how to calculate distances using shadows. For example, the distance between two obelisks of the same height is proportional to the difference in the length of their shadows. The Babylonians, too, knew the ancient technique of the shadow rod. Nevertheless, neither the Babylonians nor the Egyptians seem to have developed a great interest in extraterrestrial measurements. They were both too interested in mapping the geography of their respective kingdoms to look much beyond Earth with the same precision applied to their lands. Over time, the knowledge of the ancient astronomers fell into obscurity, rediscovered again by the Greeks. Thales of Miletus, the father of Greek geometry, who visited Egypt in the 6th century B.C.E., is reputed to have astounded the Egyptians when he correctly computed the height of the Great Pyramid by measuring the pyramid's shadow at the exact same time that his own shadow was equal to his height.

We know from the Rosetta Stone that the Greeks some time before the 2nd century B.C.E. were able to translate the ancient hieroglyphic writings. Just how much of the ancient knowledge came into their hands remains unknown. When the great library of Alexandria burned, whatever old books existed in the library's collection perished. Alexander in his conquest of the world had seen it necessary in order to consolidate his power to dismantle the great capitols of ancient Egypt at Heliopolis and of the Persians at Persepolis, killing the priests and allegedly sacking the great libraries. Heliopolis, *On* of the Bible, was once reputed to be the greatest university in the world with 13,000 priests where Moses in the time of Ramses is said to have studied the wisdom of the Egyptians. Literally, the city of the Sun, the great courtyard at Helios was made of polished black stone, so finely polished that it reflected the stars, as if one were walking among them on Earth. It stands unproved that the Greeks copied the Egyptians or the Babylonians. However, they made use of old Egyptian observatories, such as the ones at Syene and Elephantine. Whether their work was original or not, it is the Greeks and not the Egyptians who we credit with the earliest measurements of the Sun and Moon, because we do not have Egyptian or Babylonian records to claim otherwise.

98

Stellar parallax

The ancient shadow rod technique of Eratosthenes and the eclipse of Aristarchus evolved into the technique for measurement of the stars called stellar parallax. In the modern stellar parallax method, the measurement of the distance between two known points, combined with the angles from each of the two points to a star, enables the distance to the star to be calculated using trigonometric functions.

The application of the parallax method to the measurement of very distant objects like stars presents a unique challenge. First, points on Earth are too close together relative to the great distances of stars to obtain reliable angle readings. Aristarchus faced this same problem in trying to measure Earth's angle with the Sun. The angles produced by such vantage points are two small to obtain accurate calculations to objects as far away as stars. However, by taking angular sightings at different points in the orbit of Earth around the Sun, satisfactory predictions are possible for distances to relatively nearby stars.

Two angular sightings taken at points that are a known distance apart, such as at opposite points on Earth's orbit and relative to an object in the background that is more distant than the one being surveyed, such as a distant star or galaxy, will result in an angular shift of the survey object against its background. Figure **10.9** illustrates this method of distance calculation. Earth's orbit around the Sun results in two opposite vantage points lettered *a* and *b* in the diagram, from which optimum angular measurements may be taken. When we know two angles of a triangle, the third we also know because all triangles contain a total of 180 degrees. Because Earth's distance from the Sun defines known

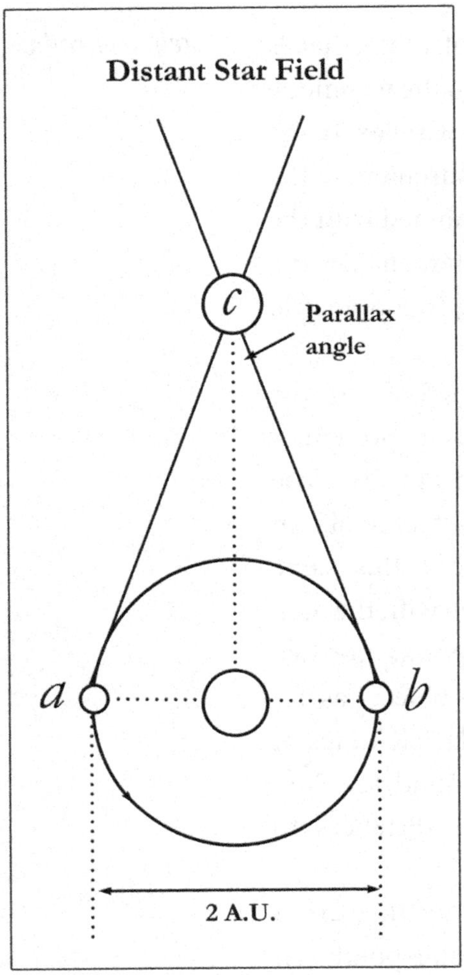

10.9. Parallax angle *c* measured from two locations on earth's orbit at points *a* and *b*.

opposite vantage points, it is possible to calculate the distance to point *c*.

We may visually demonstrate the phenomenon of a parallax angular shift by viewing a stationary object held at an arm's length. By closing first one eye, then the other, the object will appear to move relative to some distant fixed point. Moving the object closer will produce a larger relative shift. In this case, the space between the eyes becomes the fixed measurement from which the angular displacements are calculated and by which we may predict the distance to the object.

Tycho Brahe, the great Danish astronomer, attempted the first parallax measurements of the stars in the 16th century and observed no apparent angular shift. The limitations of his equipment prevented him from being successful, but his work provided the earliest speculations about the immense size of the universe. The fact that Tycho observed no angular displacements suggested that the stars were so far away that no parallax was detectable. In addition, although parallax did finally produce successful measurements made in the 19th century, it was not until the 20th century that astronomers used additional measurement techniques to obtain a hint of just how large the universe really is.

99

Parsec

In celestial measurement, the angle of parallax displacement is a *parsec* or parallax second. By definition, a parsec is one second of the arc of a circle, 1/3600 of a degree. There being 360 degrees in a circle, parsecs are very fine measurements, and hence there is considerable uncertainty in small parallax angles for objects at very great distances. With Earth as the only vantage point from which to aim, it is difficult to obtain reliable parallax angles. Significantly close parallax measurements occur for only about one in every 5,000 stars, even with measurements taken over a number of years from different points on Earth's orbit.

One parsec is equal to 206,265 Astronomical Units (AU). An Astronomical Unit is the mean distance from Earth to the Sun, or about 150 million kilometers. A parsec, then, is equal to 206,265 x 150 million kilometers, or about 31 trillion kilometers. Correlated units of measurement for parsecs, kilometers, miles, and light-years are in **10.10.** To find equivalent measurements in kilometers or miles, locate the desired unit on the light-year scale, then read vertically to the reference line, and then horizontally left or right to determine kilometers or miles, respectively. To find light-years when kilometers or miles are the known variable, reverse the procedure. The graph also provides a scale of parsec equivalent measurements as an alternative to the light-year scale used in the same way when parsecs are a known variable. For example, one parsec equals 3.2616 light-years, or in *scientific notation* 3.0857E+13 kilometers, an extremely large number. Most stars and galaxies are located beyond 100 light-years. Therefore, to accommodate distances greater than 100 light-years, multiply the scale in **10.10** by the appropriate factor. For example, to obtain measurement equivalents for objects between 100 and 1,000 light-years away, multiple the scale in **10.10** by a factor of 10.

On the rare occasions when measurements of the universe appear expressed in kilometers instead of light-years or parsecs, the numbers used often run into the trillions. A trillion is a one followed by 12 zeros. One may

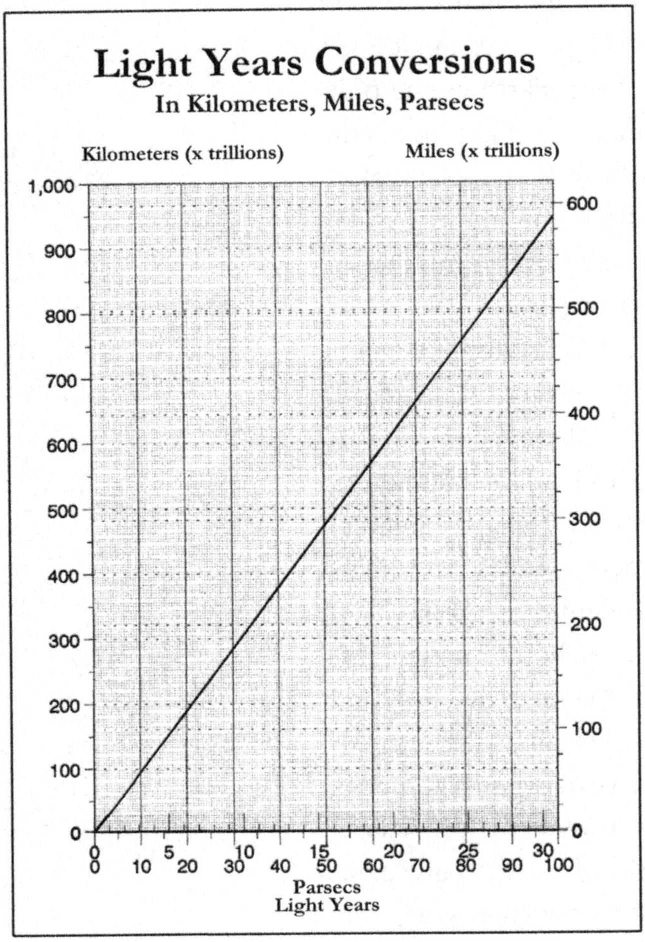

10.10. Distance Equivalents in light years, kilometers, miles, and parsecs.

visualize the scale of a trillion of something by a walk to the Sun and back. Although the Sun is at a great distance, it would take less than a half trillion steps to complete the journey. In another example, counting to one trillion very fast, non-stop every day would require eight to ten thousand years depending on how fast one could count. In another example, in just two hundred years the world's population grew from one-billion to six-billion people. At that growth rate, Earth's population could reach a trillion people in about 1,500 years, much faster than one can count. If a trillion people lined up ranked by age and marched into the sea ten abreast every minute, the population would not be exhausted. New generations would be born and grow to adulthood before the march to extermination reached their place in line. One trillion is a very large number, and our closest star neighbor is 40-trillion kilometers away.

100

Scientific notation of power numbers

Consideration of the expanse of the universe necessarily involves large-scale measurements. The large numbers ordinarily used to describe cosmic distances may appear irrelevant to terrestrial existence, until one realizes that equally large numbers describe the microcosmic world as well. For example, there are 25 trillion red blood cells in the human body. Moreover, estimates are that there are as many atoms in a single molecule of DNA as there are stars in a typical galaxy.

When encountering very large numbers or very small fractions, it is convenient to employ shorthand methods of mathematical calculation. Scientific notation is a short hand method of stating numbers that otherwise would be difficult to read. Since the study of light and other questions of the universe involve extremes of both large and small measurements, such shortcuts are particularly useful. One commonly applied method of calculation is scientific notation of power numbers. Scientific notation is the expression of a number whose base is 10, usually raised to some power. For example, 10^2 (10 squared) is ten to the power of two, or the square of 10 which is 100. The number 10^3 is the cube of 10, or 1000, (10 x 10 x 10 = 1000); the number 10^4 is 10,000 in digital form, etc. The number 3 x 10^4 is scientific notation for 30,000, or 3 x 10,000.

Sometimes the notation may look like this: 3 x 10^{-2}. The -2 (minus 2) exponent indicates this is reciprocal notation. That is, the -2 exponent in this case indicates the fraction $1/10^2$, which is equal to 1/100. Therefore, 3 x 1/100 = 3/100 =.03 in decimal form. This procedure applies to any reciprocal exponent.

It may be evident from the above examples that scientific notation easily converts from scientific to digital form and vice versa by simply locating the decimal point and counting the number of places to the right or left of the decimal indicated by the exponent. The exponent will always give the number of places the decimal moves from unity. In the example 3.0 x 10^{-2} the decimal is behind the three because it is a whole number. The exponent is -2 (minus 2), so count left two places to arrive at .03. Had the exponent been positive, i.e., 3.0 x 10^2, count from the decimal behind the 3 to the right two places to obtain 300. The same process works in reverse as well. The number 5,280 in digital form becomes 5.280 x 10^3 in scientific notation.

The usefulness of this method may not be readily apparent from these few examples. However, in the study of cosmic distances or of light and the energy associated with it, it is common to encounter notations above the order of 10^{30} or 10^{-24}. Obviously, if written in common digital form, these numbers can be lengthy and cumbersome. This can be especially annoying in mathematical complex applications like division and multiplication where one can become easily lost in long strings of numbers.

To multiply two notation numbers, multiply their multiplicands and add the exponents of their bases. To divide notation numbers, divide the dividend by the divisor and subtract the exponents of the bases. Examples:

Multiply 4 x 10^{30} by 2 x 10^{-24}

$$\begin{array}{r} 4 \times 10^{30} \\ \underline{2 \times 10^{-24}} \\ 8 \times 10^{6} \end{array}$$

First, multiply four by two to equal eight; add the exponents, or in this case subtract since the exponent of -24 is a negative number, to equal six. The answer is 8 x 10^{6}, or in digital form 8,000,000.

Divide 4 x 10^{30} by 2 x 10^{-24}

$$\begin{array}{r} 4 \times 10^{30} \\ \underline{2 \times 10^{-24}} \\ 2 \times 10^{54} \end{array}$$

Divide four by two to obtain two. This time subtract the exponents 30 -(-24) = 30 + 24 = 54 because -(-24) becomes +24 in simple algebraic logic. Now the answer is 2 x 10^{54}, a number digitally equal to two followed by 54 zeros.

Similarly, divide 2×10^{-24} by 4×10^{30}. Here we obtain the number $.5 \times 10^{-54}$, or five preceded by 54 zeros because the decimal is in front of the five and we count from there to the left 54 places.

These are the simplest forms of scientific notation. Other variations exist. For example, in place of the familiar 10^2 and 10^{-2} base and exponent form, the substituted forms of 1E+2 and 1E-2 may replace them in computerized applications of scientific notations. This avoids the cumbersome superscript typeface. The *E* assumes the position of the base 10; 10^2, for example, is 1E+2. The statement $3 \times 10^2 = 300$ is the same as 3E+2 = 300. Conversely, 3E+4 becomes 3×10^4 in base ten notation, the digital number 30,000. Division and multiplication of computer notation is the same as traditional scientific notation already described.

101

The most reliable measurements of stars using the stellar parallax method have occurred since 1960. To accomplish these dimensions, it was necessary to expand the distance between points on Earth's orbit by the application of satellite technology. The orbit of a satellite moves the vantage points for measurement beyond the confines of Earth's orbit. Just as the advent of more powerful telescopes in the early 19th century first made parallax feasible, new technologies, like satellites, are now pushing the boundaries of space even further out.

ESA Hipparcos

Two of the most important satellites launched for space exploration in the 20th century were the Hubble Space Telescope and Hipparcos. These two satellites, launched into space in the latter part of the 20th century, greatly enlarged the understanding of the universe.

A landmark achievement of modern science, the Hubble Space Telescope is roughly the size of a school bus (**10.11**). The Hubble orbits Earth at a distance of 560 kilometers, circling Earth every 97 minutes. Sent up by the United States National Aeronautics and Space Administration in 1990, the Hubble has surveyed thousands of distant cosmic objects and relayed to Earth photographs of deep space that comprise some of the most striking pictures of the universe ever recorded.

Less notable perhaps was the satellite Hipparcos. Hipparcos had a different mission from the Hubble. The European Space Agency launched Hipparcos in 1989, and for five years it measured parallax angles a hundredfold better than any previous parallax data.

10.11. The Hubble Space Telescope.

Hipparcos fixed distances, motions, and directions of stars with an accuracy previously unknown. It followed an elliptical orbit that ranged to 37,000 kilometers. This 1.4-ton satellite examined more than 100,000 stars before ending its mission in 1993. Its name, intended as a tribute to the Greek astronomer Hipparcus, was also an acronym for High Precision Parallax Collecting Satellite.

From their respective vantage points in space outside Earth's atmosphere, the instruments of the Hubble Space Telescope and Hipparcos gathered information that caused a rewriting of astronomy and opened the universe to a completely new set of questions about its origin, size, and age.

The ESA Hipparcos Catalog came out in 1997. Published based on the new information Hipparcos produced, it caused reassessment of opinions about the universe. For example, Hipparcos detected a warp in the shape of the Milky Way Galaxy, possibly due to invisible Dark Matter in the halo around the galaxy. At the same time, the probe discovered that the Pleiades star cluster, a major reference in helping to determine the size of the universe, is 10 percent closer than originally thought. Similarly, the Hyades, another important reference star cluster, underwent reevaluation under the new findings. Because of the Hipparcos reassessment, the entire scale of the estimated size and age of the universe changed.

While Hipparcos revealed that some stars are closer, others it found to be at greater distances than once predicted. For instance, the star identified as HIP48152 is twice as far as originally thought. Distance checks on many stars produced similar underestimates. Analysis of the new data called into question existing estimates of star ages and the dynamics of the evolution of stars. If the Hipparcos discoveries hold up under future scrutiny, the size of the universe could increase by more than 10 percent.

Prior to Hipparcos, uncertainty in the traditional parallax method of stellar measurement began to emerge at about 65 light-years, and the method was generally unreliable beyond 100 light-years. Even with the Hipparcos satellite and the most sophisticated modern instruments available, stellar parallax as a celestial surveying technique still is not practical beyond about 300 light-years, certainly not the millions of light-years to the galaxies. The farther away the star, the smaller and consequently less certain the parallax angle of measurement is, until beyond a certain point parallax measurement is not feasible regardless of the vantage point, and other methods of predicting distances must then be used.

Southern Ring Planetary Nebula. Shaped like an irregular oval with lacy, reddish orange plumes of gas and dust, the inside of the circle glows bright blue. In the center of the rings are two stars, one glowing much brighter. Stars seen outside the nebula lie at vastly different distances. James Webb Space Telescope. NASA, ESA, CSA, and STScI.

XII

SPHERICAL GEOMETRY AND MAGNITUDE OF STARS

102

Star magnitude is a means of estimating stellar distance. In the wide range of brightness that covers all stars, the brightness of a typical star is about the same as our Sun. Thus, by applying a standard mathematical treatment known as the inverse square law to the apparent magnitude of a star, we may predict the star's distance relative to the known distance to the Sun and the Sun's magnitude. However, since all stars are not quite like our Sun and are not always the same brightness, this method, as in stellar parallax, is subject to inaccuracy, and it requires knowledge about individual stars in order to interpret correctly their magnitudes. This knowledge we express in the assignment of each star to its place on the Absolute Magnitude Scale, which is an internationally standardized reference that has evolved as one of the major tools of astronomy.

103

Hipparchus

The magnitude scale of the brightness of stars is rooted in history. The Greek astronomer Hipparchus was the first to catalog systematically the magnitude of the stars in the 2nd century B.C.E. Of all the early astronomers, Hipparchus was among the most accomplished. Many historians consider him to have been the greatest of the Greek astronomers. Best known for his star catalog, he also produced, in addition to calculations of celestial distances, predictions of Earth and lunar motions, predictions that turned out to be more accurate than those of many astronomers who followed him were. His observations and measurements of the stars and planets were the most careful and accurate of his time. He calculated the length of the solar year to within 6.5 minutes, and discovered the annual *precession* of the equinoxes to be 46 *seconds of arc*, the best calculation until modern times. Precession is the complex movement of Earth on its axis. Sometimes described as a wobble, the movement causes Earth to cross the circle of the ecliptic path each year slightly ahead of its crossing the previous year. Seconds-of-arc refers to a tiny piece of the 360 degrees of the ecliptic circle.

Of his many accomplishments, perhaps the greatest contribution of Hipparchus to astronomy and to mathematics, besides his star catalog, was his invention of trigonometry, the method of relating the angles and lengths of sides of right triangles. He adapted the ancient Babylonian circle of 360 degrees, and his original tables of trigonometric chords are very close to those that appear in modern geometry texts. Hipparchus did not use the modern trigonometric expressions of sine, cosine, and tangent; these came later, but rather applied Euclidean geometry theory to relate the different ratios in his tables to the 360-degree circle. Hipparchus' trigonometry became, and still is, a mainstay of stellar distance measurement.

Hipparchus himself posted distance estimates to the Moon and Sun, as well as measuring the lunar orbit and what he perceived at that time, in Aristotelian fashion, to be the solar orbit around Earth. He was also the first to describe an eccentric, non-circular path for these orbits, and the first mathematically to describe solar eclipses.

Born in Nicaea (now Iznik), Bithynia (now Turkey), Hipparchus was at Alexandria at the height of that great city's intellectual influence. His life spanned only about

55 years. He was born around 180 B.C.E. (some scholars place his birth as early as 190 B.C.E.) and died in Rhodes, Greece, around 125 B.C.E., remembered by the Greek world as Hipparchus of Rhodes. We know little of Hipparchus' life. His work appears to have been one of the foundations for the theories of Ptolemy three hundred years later who wrote about his Greek predecessor. Hipparchus rejected earlier Greek theories for a heliocentric solar system, instead choosing to develop a complex system of epicycles that became a major tenet in Ptolemy's description of the Sun and planets moving around Earth.

The confounding nature of the universe, even to the genius of Hipparchus, is in his calculation of the distance to the Moon. Like Aristarchus before him, he achieved his measurement with uncanny accuracy by comparing the position of the Moon to that of the Sun, which he set at infinity. For centuries afterward, the relationship between Earth and Sun was to occupy center stage in astronomy, because no one knew that the stars were actually farther away than the Sun.

104

The magnitude scale

The Hipparchus star catalog of 850 stars, which he compiled over a period of several years, was of such high quality that it was in use from its date of completion in 129 B.C.E. until well after 1677 when the English astronomer, Edmund Halley, referred to it in his own observations.

Hipparchus designated the brightest stars in his catalog as being of the first magnitude. Proportionately dimmer stars he listed as magnitude 2, 3, etc., down to magnitude 6. Magnitude 6 stars are the faintest stars visible with the unaided eye. Ptolemy improved this visual magnitude system in the 2nd century. Nevertheless, for nearly two thousand years the magnitude system relied primarily on Hipparchus' visual estimation of brightness.

An early source of error in trying to determine the magnitude of a distant star was the lack of reliable measurements of its brightness. Early astronomers had no photometric processes for establishing brightness. Theirs was purely a visual estimation.

This problem of erroneous magnitude resurfaced in the work of Johann Bayer. Astronomy books still reprint his 17th century constellation charts although his charts

11.1. The Constellation Orion, from Johann Bayer's *Uranometria*, first published in 1603. Bayer drew his map backwards as if looking in from outside the celestial sphere. United States Naval Observatory Library.

were inherently wrong in several instances (**11.1**). Bayer introduced the idea of identifying stars with Greek letters in the order of their magnitude within a constellation; e.g., Alpha Ursa Major is the brightest star in the constellation Ursa Major, Beta Ursa Major is the second brightest, and so on. The difficulty with Bayer's assignment of magnitude was that he sometimes incorrectly assigned these letter designations. For example, we know today that Alpha Orion is actually dimmer than Beta Orion. Bayer's work appeared based in part on the star catalog of Hipparcus, known itself for misjudgments of the magnitudes of stars. For other reasons, though, Bayer may not be at fault, because Alpha Orion actually may have been brighter in Bayer's time. Alpha Orion, known also by the name Betelgeuse, the red supergiant star is now in the latter phases of its stellar life.

Meanwhile, the stars Alpha and Beta Andromeda, two other examples in Bayer's charts, which should be brighter and dimmer respectively, are of equal visual magnitude. All of this is to say that early efforts to measure stellar distance by magnitude were subject to much uncertainty.

With the advent of photography in the mid-19th century, it became possible to arrive at a more accurate mathematical basis for the magnitude scale. Calculations made from timed exposures of photographic plates gave good estimates of the amount of light required to produce images of stars in different magnitude classes. This system and its subsequent improvements in cameras and photometric technology was the mainstay of astronomers in mapping the universe for more than a hundred years.

11-2. Photograph of the constellation Orion showing the apparent brightness of stars recorded by exposure to film.

The image in Figure **11.2** illustrates the application of photography to the magnitude scale. In the figure, each dot represents a celestial object of a certain magnitude. In photographs of the sky, the uniform sensitivity of the film reacts to the varying brightness of the different "stars" to produce a larger dot for brighter magnitude stars and systematically smaller dots for dimmer magnitude stars. The three stars of Orion's belt appear at the center of the photograph.

105

Early photometric measurements revealed that a difference of five magnitudes on the Hipparchus scale amounted to a measured difference of about 100 times in brightness. Subsequent adjustments corrected the magnitude scale, so that the present day scale corresponds exactly to a factor of 100 from magnitude +1 to +6; that is, magnitude +1 is 100 times brighter than magnitude +6. The historical tradition of the magnitude scale remains intact; lower magnitudes receive higher numbers, while brighter stars have the lesser numbers on the magnitude scale. (**11.3**)

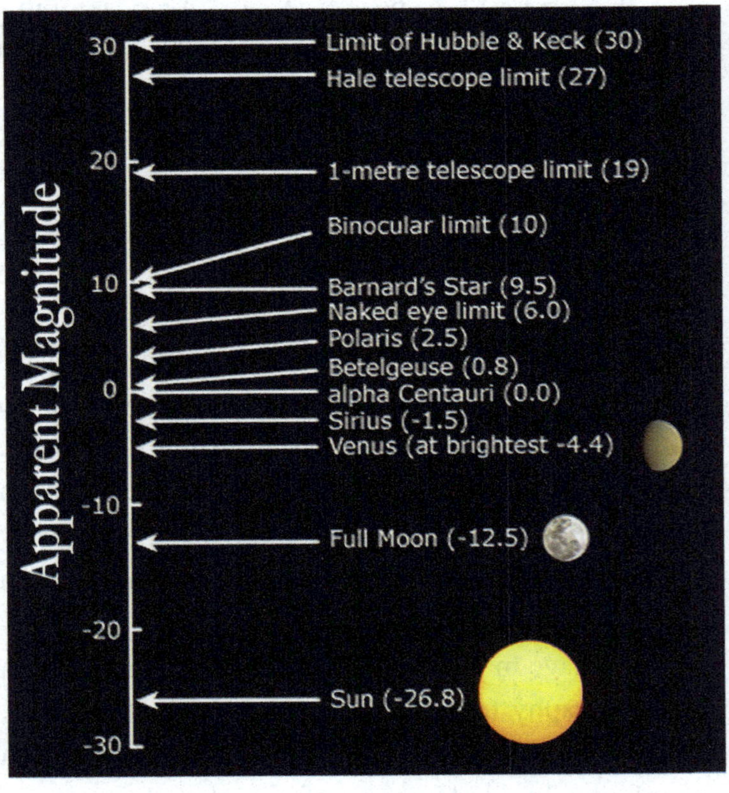

11.3. Apparent magnitude scale. The naked-eye limit to see a star is 6.0 on the Apparent magnitude scale. European Space Agency.

To determine the brightness factor at each magnitude step, one needs only to know the fifth root of 100, since there are five magnitude steps increasing from +1 progressively up to +6. The fifth root of 100 is 2.512. Therefore, each magnitude step increases or decreases in brightness by a factor of 2.512. Magnitude +2, for example, is 2.512 times less bright than magnitude +1; magnitude +3 is $(2.512)^2$, or 2.512 x 2.512, or 6.31, times fainter than magnitude +1; magnitude +4 is $(2.512)^3$, or 2.512 x 2.512 x 2.512, or 15.85, times fainter than magnitude +1, and so on. The unaided eye is capable of detecting magnitudes to about magnitude +6, but the strongest space-based telescopes can detect magnitudes above +30. The Hubble Space Telescope, for example, can see objects one-billionth as bright as the human eye can see. At the other end of the brightness scale, the Sun comes in at about magnitude -27, by far the brightest object in the sky. The new Webb Space Telescope promises to detect objects above +34 magnitude.

When analysts applied photometric measurements to the original Hipparchus scale, they found a few stars were actually brighter than magnitude +1. Some were at zero magnitude on the scale, while others were even brighter. Having adopted a standard factor of brightness change of 2.512, it was then a simple matter to extend the scale into the minus range for the brightest objects in the sky. Sirius, for instance, is the brightest star in the sky after the Sun at magnitude -1.4. The Sun and Moon by comparison are -26.8 and -12.7 respectively. The planets Venus, Jupiter, and Mars come in at -4.0, -2.5, and -2.0 in that order. The planets are brighter than any of the stars are because they are much closer.

Similarly, a very faint object may receive a magnitude rating beyond the magnitude +6 limits of the unaided eye. A medium sized telescope, for example, will make visible celestial objects as faint as about +12 magnitude, which is 39 million times fainter than magnitude +1.

The old system of using a photographic plate exposure was good for measuring magnitudes down to around +17. After 1950, sophisticated electronic means pushed the magnitude scale beyond +24. The successful deployment of the Hubble Space Telescope permitted the detection of objects with brightness as faint as magnitude +31.5. Thus, from a range of 100 brightness differences spanning six magnitudes between the faintest and brightest observable objects under the ancient Greek scale, the system today spans more than 30 magnitudes — not including the Sun — and it seems to expand with each new technological advancement.

Thirty steps of magnitude represent a brightness difference of 2.5 trillion times from the brightest to the dimmest object represented. If we include the Sun at -26.8 magnitude, then the brightness difference from the Sun to the faintest detectable object in the universe approaches 60 magnitudes. Discounting the use of the telescope, the difference is still more than 10 trillion times from the faintest star visible to the unaided eye to the brilliance of the Sun. The visual system will not accommodate direct viewing of the Sun. Yet the incredible adaptive capacity of the eye that makes possible the processing of such a fantastic range of light stimuli seen in the stars begins to come into appreciation.

106

Some astronomers find the magnitude scale to be outmoded, preferring instead to use flux values of distant stars and galaxies. Flux is the luminous light energy that the visual process transforms into the brightness sensation. Flux is a way of bypassing the visual system in favor of a purely mechanical measurement of energy. The Sun, for example, has an observed flux value of 1.39

kilowatts per square meter. This one square meter of the emitted energy of the Sun's surface, measured at Earth's orbital distance, we know as the solar constant. The solar constant is not to be confused with solar luminosity, which is a measure of the entire solar disk. The Sun's luminosity, as inferred from the solar constant, is about 3.9E+23 kilowatts.

Each star has its own luminous flux value. From such data, it is possible to estimate star luminosities and other characteristics such as distance in the same way that magnitudes are used but without the visual magnitude scale. The magnitude scale functions as a set of disguised flux values. However, it is unlikely that the magnitude scale, a system that has stood for more than two thousand years, will disappear easily in favor of the flux system, particularly when the progression of brightness perception embodied in the visual magnitude scale so naturally fits to the human perceptual mechanism.

107

Apparent and absolute magnitude

The magnitude scale is subject to interpretation in two forms: apparent and absolute. Apparent magnitude is the observed brightness of a stellar object as viewed from Earth, while absolute magnitude is a calculated brightness for such a body if viewed from a set distance, a distance standardized at 10 parsecs, or about 310 trillion kilometers. For example, the Sun if viewed from a distance of 310 trillion kilometers instead of its known 150 million kilometers would have a magnitude of just +4.79. At that distance, it would appear to the naked eye to be dimmer than many stars in the sky. It would be visible, but barely so.

A list of the eleven brightest stars and their apparent and absolute magnitudes is included in Table **11.4**. These eleven stars are visible with the unaided eye, and all but two are within 100 light-years of Earth. The table does not include Rigel or Betelgeuse, bright variable stars that vary in brightness over a period. In the table, apparent magnitude is the brightness of a star as seen from Earth, while absolute magnitude is the brightness of the same star at a set distance of 32.6 light-years (10 parsecs). The lower the number, the brighter the star is. Stars with negative numbers are the brightest.

		Constellation		Magnitude		Distance	
	Proper name	Greek/Latin	Translated	Apparent	Absolute	Parsecs	Lt. Yrs.
0	Sun	Sol	Sunne	-26.8	+4.79	0.0	0.0
1	Sirius	*alpha* Canis Majoris	Great Dog	-1.45	+1.4	2.7	8.7
2	Canopus	*alpha* Carinae	Ship's Keel	-0.73	-4.7	60.0	98.0
3	Rigil Kentaurus	*alpha* Centauri	Centaur	-0.1	+4.3	1.3	4.3
4	Arcturus	*alpha* Boötis	Herdsman	-0.06	-0.2	11.0	36.0
5	Vega	*alpha* Lyrae	Harp	+0.04	+0.5	8.1	26.5
6	Capella	*alpha* Aurigae	Charioteer	+0.08	-0.6	14.0	45.0
7	Procyon	*alpha* Canis Minoris	Little Dog	+0.35	+2.7	3.5	11.3
8	Achernar	*alpha* Eridani	River Eridanus	+0.48	-2.2	39.0	127.0
9	Hadar	*beta* Centauri	Centaur	+0.60	-3.5	150.0	490.0
10	Altair	*alpha* Aguilae	Eagle	+0.77	+2.3	5.0	16.5
11	Aldebaran	*alpha* Tauri	Bull	+0.86	-0.2	16.0	68.0

11.4. The eleven brightest stars below magnitude 1, listed in order of apparent magnitude.

Stars have names. Stars in a constellation we know by their proper names or by their Greek and Latin names, as Bayer devised, according to their apparent magnitude as alpha, beta, gamma, etc.; Beta Centauri, for example, is the second brightest star in the constellation Centaurus after the star Alpha Centauri. In addition, stars may have other names. Alpha Centauri, for example, retains its medieval Arabic name, Rigil Kentaurus. We may know other stars by numbers according to their listings in star catalogs, or in various other listings of cosmic objects.

Just as we calculate the absolute magnitude of the Sun by applying the inverse square law to the Sun's apparent magnitude of -26.8, a similar interpolation technique applies to estimate distances to other stars, except in reverse. If the magnitude of a star is known, then its distance may be calculated because we know the distance-magnitude relationship of the Sun and of nearby stars through the stellar parallax method of estimating distance. To measure distant stars beyond stellar parallax range, it is necessary only to know their absolute magnitudes, that is, how bright they appear from the standard distance of 10 parsecs. Nearby stars may appear brighter than very distant ones although they may actually radiate only a fraction as much light. For example, the star Alpha Ursa Minor, Polaris the North Star — the brightest star in the constellation of the Little Dipper — is 200 parsecs from Earth but appears to be nearly as bright as the star Dubhe, the brightest star in Ursa Major, the Big Dipper, which is by comparison 30 parsecs away.

108

The Inverse Square Law

The inverse square law is a natural phenomenon that describes the relationship between a point of origin and the surface of a progressively enlarging sphere (**11.5**). Simply stated, the inverse square law says that the area inscribed on a sphere varies inversely proportional to the distance of the surface of the sphere from its point of origin, or center. One unit of area coverage at one centimeter, for example, will expand to four units of surface coverage at two centimeters, nine at 3 centimeters, 16 at four, etc., thus maintaining the same proportional coverage of the surface of the sphere at any radius. The intensity of a star at twice its known distance would be 1/4 as great. In such a case, the square of the distance factor of two is four. When stated inversely, four becomes 1/4; thus, the intensity is 1/4 of the original amount resulting in an observed brightness that is four times fainter just by doubling the distance.

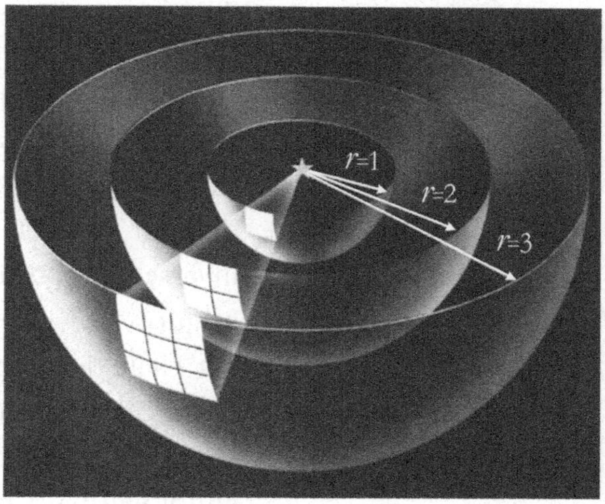

11.5. The inverse square law of spherical propagation. J. Pasachoff and M. Kutner, *University Astronomy*.

Radiant energy spreads out from its center of origin in an ever-enlarging pattern. At two times its unit distance, the energy must cover four times as much area; at three times the distance, the pattern has enlarged to nine times in area. The amount of light on each unit of surface diminishes accordingly. At three times the unit distance, the radiated energy is only one-ninth as great as at its source. Thus, the amount of energy arriving at any one location is inversely proportional to its distance from its origin.

Spherical expansion creates a natural dissolution of spherical radiation as a function of the radius of the sphere. Radiance concentrated on one unit of a spherical surface at one radius unit of distance must cover four units of the surface at two radii. The original single unit of coverage spreads over four units of a spherical surface instead of one unit; hence, each of the four units is 1/4 as intense as the coverage furnished by the original one unit.

109

All point sources of any spherical expansion, or radiation, and/or attraction, as in the case of gravity, are subject to the inverse square law. It is a well-known fact that all radiant energy diminishes in its intensity inversely proportional to its distance from its point of origin.

There is nothing particularly magical about this. Spherical expansion from a center of origin will always have predictable radiance characteristics. Any spherical entity will have properties that conform to the law of inverse squares. Newton's law of gravitation is another prominent example of such spherical geometry. The force of gravitational attraction varies between two masses inversely as the square of the distance between the objects. The force between two objects at one unit of distance is 1/4 that amount at two units. In this case, the sphere of gravity behaves as if the mass were synonymous with its center of origin. This phenomenon recalls the presumption that the cosmos evolved as matter that migrated toward gravity points, and eventually transformed into mass. Each original point of material is its own collector of mass. Set against the delicate balance of other gravity points throughout the universe, each spatially distinct mass has its own conformation to spherical geometry.

110

The Chéseaux equation

One of the earliest successful attempts to apply the inverse square law to show that the magnitude of the stars signaled their great distances was by the 18th

11.6. Portrait of Jean-Philippe Löys de Cheseaux. Painting by J. P. Huechez. Cantonal Museum of Fine Arts, Lausanne, Switzerland.

century Swiss astronomer Jean-Philippe de Chéseaux, pictured in **11.6**. Chéseaux's ingenious argument still draws the praise of modern astronomers.

Chéseaux was the first to put a number to the speculation about stellar distance and to back up his claims with scientific evidence. His method was mathematically sound. For the first time, stars appeared to lie at far greater distances than ever believed before. Others who preceded Chéseaux, like Tycho Brahe and Johannes Kepler, had theorized about the distances to the stars, but their theories — even though in those instances coming from renowned astronomers — were no more proven than were those of Ptolemy's day. Chéseaux proved that stars lie far off. The Chéseaux discovery was to eighteenth century astronomy what Aristarchus' was to antiquity and what Shapley and Hubble's discoveries about the size of the Milky Way Galaxy and the universe were to 20th century astronomy.

A few simple mathematical symbols help to illustrate the cleverness of Chéseaux's method. From stellar parallax measurement, Chéseaux knew the size of Mars' orbit, its distance from the Sun being on average approximately 228 million kilometers, represented by ***R*** in **11.7** and **11.8**. In addition, he also knew from parallax measurement in the 18th century that the distance from Earth to the planet Mars (capital letter ***D***), calculated to be about 312 million kilometers. In the configuration used by Chéseaux and illustrated in **11.7**, Chéseaux

11.7. Variations in the Mars orbit. Chéseaux used the Sunlight reflected from Mars to estimate the distance to the stars.

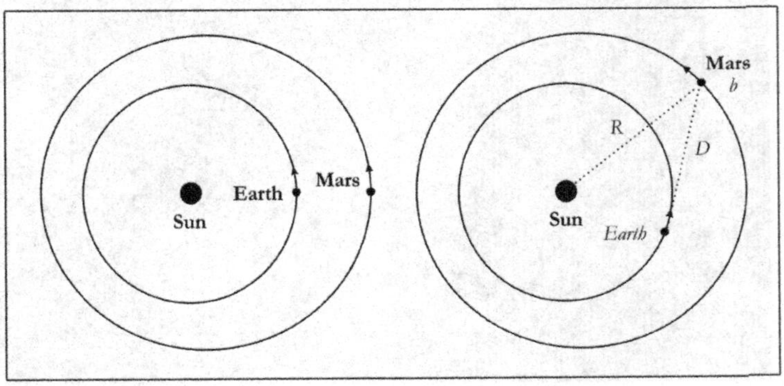

chose a configuration in which Mars was at its farthest distance from Earth, yet still visible in the evening sky. At one point in its orbit, Mars is about five times farther from Earth than at its closest location, and is roughly 15 times dimmer.

Chéseaux likewise knew the radius of Mars, at the time estimated at 3,380 kilometers (***b*** in **11.7**). What Chéseaux did not know and had no way of measuring was the solar energy of the Sun or the luminosity of any of the distant stars, nor did he have any way to tell how far away the most distant stars were. In **11.8**, ***L*** stands for the unknown luminosity in the equation and ***d*** is the unknown distance.

$$\frac{L}{4\pi d^2} = (\pi b^2)\cdot\left(\frac{L}{4\pi R^2}\right)\cdot\left(\frac{1}{4\pi D^2}\right)$$

$$\vdots$$

$$d = \frac{2R\cdot D}{b}$$

$$d = \frac{2\cdot 228{,}000{,}000\cdot 312{,}000{,}000}{3380}$$

$$d = 42 \textit{ trillion kilometers}$$

11.8. The Cheseaux Equation.

Mars, at its nearest point to Earth, has an apparent magnitude of about -2, but at its most-distant point, it is of about the first magnitude. Applying the inverse square law, Chéseaux was able to calculate the amount of solar energy reflecting from Mars as a function of the size of the Mars disk, based on its known radius, and the distance of Mars from the Sun when it was at its farthermost, first-magnitude point in the sky. While this did not produce an amount of luminous energy, it did produce a luminous quantity for Mars relative to the Sun at a known distance from Earth. For a star of the same luminous output as the Sun to appear of equal magnitude to Mars, it would have to be at distance ***d***. The genius of the Chéseaux equation was that when the various quantities were arranged in formula, luminosity

— one of the unknowns in the equation — canceled out, leaving the relatively simple equation $d=2R \cdot D/b$, where everything is known except d, or distance. In mathematics when the same terms appear equally on either side of an equation, as in the case of luminosity L, those terms cancel.

Solving in the equation $d=2R \cdot D/b$, using quantities known to Chéseaux, d is found to be about 42 trillion kilometers, or around 4.4 light-years. Alpha Centauri, the nearest star to Earth, is in fact 4.3 light-years away. Chéseaux's estimate of a few light-years distance to the stars proved to be in good agreement with what would turn out to be the actual distance to the nearest star.

Chéseaux made two assumptions in setting up his equation: all stars have the same luminous energy as the Sun, and Mars reflects 100 percent of Sunlight. Both of these assumptions were incorrect. The reflectance of Mars is 15 percent instead of 100 percent, and not all first magnitude stars are like the Sun in their luminous energy. The luminous energy reflected from Mars is actually much less than 15 percent when the angle of observation shown in **11.6** comes into consideration. Correcting for Mars' reflectance, the Chéseaux equation produces a distance of more than a dozen light-years instead of the conservative estimate of 4.4. Nevertheless, in the 18th century, observers thought the stars generally to be much closer than the distance arrived at by Chéseaux. Chéseaux believed that in astronomical terms his results would be close enough to produce a credible new approximation of stellar distance. An estimate of a dozen light-years would have been unbelievable. Using incorrect data, Chéseaux arrived at a nearly correct result.

Surprisingly, Chéseaux is relatively unknown. His name rarely appears in scientific references, and then only in connection with his work on comets, specifically the Comet de Chéseaux, or the Great Comet of 1744, that carries his name. His book, *Traité de la comète qui a paru en Décembre 1743 & Janvier, Fevrier & Mars 1744*, published in 1744, described the comet's behavior and orbit, as well as other celestial phenomena, including his prediction of the distance to the stars. Apparently well respected in his own time, Chéseaux's work today lies all but forgotten. Other 18th century scientists like the great Swiss expatriate Leonhard Euler who in 1744 published his classic book on celestial mechanics overshadowed Chéseaux. Euler we remember as the most prolific mathematician in history, while Chéseaux rests in obscurity seldom mentioned.

111

Inverse square measurement

The measurement of distances using the inverse square law as applied to star magnitudes is a simple matter. If the stellar parallax method establishes the distance to a nearby star of known magnitude, such as the star Alpha Centauri, and we know the magnitude of a more distant star of the same size and type as Alpha Centauri, then we easily calculate the distance to the second star from the inverse square law. The trick in this procedure lies in knowing that the second star is comparable in size and type to the comparison star, in this case Alpha Centauri.

112

It turns out that if viewed from the standard distance of 10 parsecs, many stars would be approximately the same magnitude as Alpha Centauri or the Sun, both of which are very similar and typical celestial bodies of their type. Certainly, we know the distance to the Sun and its magnitude. Therefore, the Sun becomes a convenient basis of comparison for measuring distant bodies in the universe. There are, of course, stars both brighter and fainter than the Sun. At least 28 of the nearest stars — those within 15 light-years — are intrinsically fainter than the Sun. Only two stars within that range are brighter. Naturally, the Sun appears very much brighter because it is much nearer to us. Beyond 15 light-years there are stars brighter than the Sun, and those that are brighter are sometimes much brighter. The most brilliant stars in the sky are 10,000 to 1,000,000 times brighter than the Sun. Some of the brightest, most-distant stars have absolute magnitudes of -9.0 compared to the +4.79 of our Sun. Still, a large number of observable stars are about like the Sun.

Carina Nebula. This three-light-year-tall cosmic pinnacle is primarily dust and gas that exhibits signs of intense star-forming activity.
Hubble Space Telescope. NASA, ESA, M. Livio and the Hubble 20th Anniversary Team (STScI). 2010.

XIII

MULTIPLE STAR SYSTEMS

In the average person's lifetime, he or she will see only a few hundred "stars," and these merely as twinkling points of light in the evening sky. As point sources of light, the stars twinkle through Earth's atmosphere, while the planets, which are disks, and therefore plane sources of light, shine steadily. The number of stars in the universe and the distances between them is almost unimaginable. Yet in the experience of most people the stars will be viewed and enjoyed without any consideration or thought given to their sizes or places in the vastness of the universe. A few thousand of the billions of them in the Milky Way Galaxy constitute our total experience with stars. What actually exists beyond the visual plane is remarkably different from what we see.

Table **12-1** lists the twelve stars closest to Earth in light-years, along with their apparent and absolute

Star Name	Distance Lt. Yrs.	Absolute Magnitude	Apparent Magnitude
Proxima Centauri	4.2231	15.5	11.05 (var.)
alpha Centauri A	4.3950	4.4	-0.01
alpha Centauri B	4.3950	5.7	1.33
Barnard's Star	5.9409	13.2	9.54
Wolf 359	7.782	16.7	13.53 (var)
Lalande 21185	8.3120	10.5	7.5
Sirius A	8.6011	1.4	-1.46
Sirius B	8.6011	11.2	8.3
Luyten 726-8A	8.728	15.5	12.52 (var.)
Luyten 726-8B	8.728	13.02	16.0
Ross 154	9.6934	13.1	10.45
Ross 248	10.32	14.8	12.29

12.1. The twelve nearest stars based on the Hipparchos satellite data. Prior to Hipparchos, the stars Luyten 726-8A and B were thought to be closer than Sirius A and B. Only two stars of the twelve listed are visible with the unaided eye, namely, the double stars alpha Centauri A and B and Sirius A and B. All the other stars in the table are too faint to see. The rank of a multiple-star system is by its brightest member. In Table 12.1, variable brightness stars are marked as (var.).

magnitudes. The list excludes the Sun since it is light-*minutes* and not light-years away. The Sun at absolute magnitude +4.79 would rank third brightest of this group of 12.

14

Sirius

Outside our own solar system, the brightest object in the entire sky is Sirius, with an apparent magnitude of -1.46. Sirius has an absolute magnitude of +1.41.

Sirius (pictured in **12.2**) is 23 times as luminous as the Sun. A hot, blue-white star, Sirius has a surface temperature of 10,000 degrees centigrade nearly twice that of the Sun. Sirius is 8.7 light-years away. Seen clearly in the winter sky of the northern hemisphere, it is located in the constellation Canis Major, or Big Dog.

12.2. Photograph of the Alpha Centauri multiple star system showing Alpha Centauri A and B. McDonald Observatory.

Five-thousand years ago, Sirius signaled the beginning of the New Year and originated folklore that lives on still today. The summer solstice, the longest day of the year, has been a time of festivals since ancient times. In ancient Egypt on the solstice, Sirius and the Sun rose together to fix the beginning of the Egyptian 365-day calendar, at a time of the year that happened to coincide with the annual flooding of the Nile. Thus, the Egyptians viewed this great star with beneficial reverence, naming it the Nile Star and the Star of Isis, goddess of fertility.

The Babylonians and the Greeks, however, found more menacing associations for Sirius. Known as the Dog Star, because of its location in the constellation Canis Major, the Dog, and its proximity to the constellation of Orion the Hunter, the Babylonians viewed Sirius with trepidation. Its appearance they considered an evil sign. The Greeks, too, believed that Sirius at harvest time produced drought, disease, and all sorts of other human misery, as well as madness in dogs. Sirius is Greek literally for the *glowing one* or *scorching one*, so named for the belief that it somehow in combination with the Sun contributed to the heat of summer.

Whether the association of Sirius with madness in dogs came before or after its naming as the Dog Star is unknown. Nevertheless, the name Dog Star goes back at least to the Babylonians. Certain uncomfortable periods of the summer are referred to still as dog days, those hot, sultry periods between July and September in the northern hemisphere when nature is stagnant and inactive and when animals are thought to be more prone to rabies, or mad-dog disease. Dog-days folklore today connects less directly to the rising of Sirius; nonetheless, Sirius the Dog Star is the origin of dog days.

115

Binary star systems

Sirius appears in the evening sky as a single star, as one might imagine the Sun to appear at a great distance. However, Sirius is distinctly different from the Sun in one important way; it owes its brightness to more than one star. This is common among stars. About two-thirds of the stars in our galaxy conceivably have companion stars. Alpha Centauri, for instance, our nearest star neighbor, is also a compound star. The Sun does not have a companion star, but it is more the exception than the rule. As interesting as Sirius is for its folk legends, even more interesting is its companion, Sirius B, the first *dark matter* discovered in the universe.

As a binary star system, Sirius is two stars orbiting each other around a common barycenter, but appears to the eye as a single star. In **12.2**, the small dot below the bright star Sirius is Sirius B. Indistinguishable to the unaided eye, Sirius B is a white dwarf star of incredible density, orbiting the main star Sirius. A white dwarf is a star whose nuclear engine has gone out. It collapses on itself and eventually becomes extinguished, to exist as an invisible mass, a tablespoonful of which is so tightly packed that it weighs a ton. Sirius B, at one time the size of the Sun, now occupies a volume only three times the size of Earth, but 100,000 times more massive.

Long before it became visible to modern technology, observers predicted the existence of Sirius B. In 1844, Friedrich Bessel, the German astronomer who also was the first to measure the distance to a star by the stellar parallax method, said Sirius had a companion star. Bessel spent ten years observing the movement of Sirius. He used its eccentric path across the sky, a result of a wobble around its Barycenter, to predict the presence of Sirius B. However, it was not until 1862 that telescopes improved to the point that Sirius B emerged detected. It orbits Sirius in an elliptical orbit every 50 years at a mean distance equal to about that of the planet Uranus around the Sun. In 1915, astronomers categorized Sirius B as a true white dwarf star, thus officially acknowledging the first *dark matter* of the universe.

116

Meanwhile, turning to the multiple-star system, Alpha Centauri, in another part of the sky, Alpha Centauri is in fact not two stars but three — a triple star system. Located in the constellation of the Centaur, Alpha Centauri is visible only from the southern latitudes.

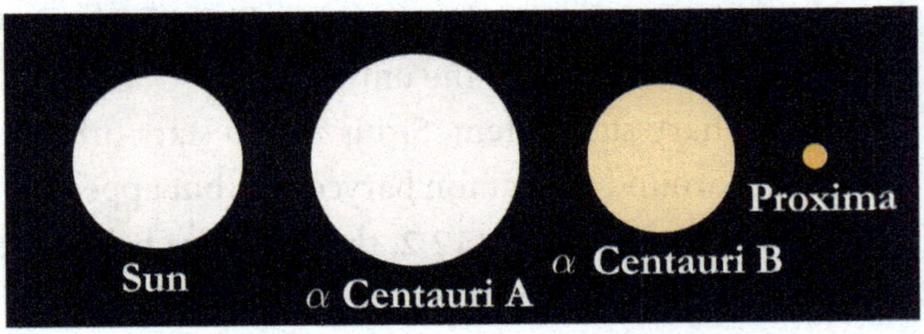

12.3. A comparison of the sizes and colors of the stars in the Alpha Centauri system with the Sun.

Each star in the Alpha Centauri system has identity simply as A, B, and C (**12.3**). Alpha Centauri A orbits

Alpha Centauri B every 80 years. Alpha Centauri C, known also as Proxima Centauri, revolves around both A and B every 500,000 years. Alpha Centauri C is a flare star. Very dim, with a luminosity 1/13,000 that of the Sun, it has barely sufficient mass to be a star, yet it produces numerous flares as great as those of the Sun, only much more frequently.

About half of the stars nearest to Earth are multiple systems, and substantially more than half of the brightest stars in the sky owe their magnitude to their compound natures. We have known the star Mizar in the handle of the Big Dipper to be a double star for centuries. The Hipparcos satellite in its remarkable five-year survey from 1989 to 1993 found no less than 24,000 double star systems, half of which existed previously unknown.

In our own solar system, the Sun, while not a double star, almost was; because the planet Jupiter came very close to becoming a star, lacking just enough atomic energy in its formation to ignite like a star. Had the Sun and Jupiter become a binary, or multiple, star system, it is interesting to speculate about our solar system. One wonders how the ancient mythologies might have reflected on such a phenomenon. Most likely the reaction to a double Sun would have become as natural as the acceptance of the Sun and Moon. More important though, than a visual or mythological reaction, would have been the climatic and environmental changes brought by such a two-star system. The planet would have been a very different Earth from the one we know today.

The universe manifests in progressively larger and more complex groupings of stars. Individual stars join as multiple-star systems, star systems group into star clusters, and clusters make up the galaxies. There are even multiple galactic systems, rounding out the universe and its macrocosmic structure.

117

Clusters of stars

Open star clusters are groups of stars that are moving through space as a group. They are important because stars in an open cluster were born at about the same time from the same gas cloud. Therefore differences in stars in a group can be used by astronomers to predict several things about types of stars, including temperature, luminosity, composition, and age.

Two popular open star clusters are the Hyades and Pleiades, both in the Taurus constellation. Located in the jaw of Taurus, the Hyades cluster is a relatively young star region believed to be less than a billion years old. The Hyades cluster rises and sets with the seasons and was expected by the Greeks to bring rain when it neared the horizon.

Meanwhile, the Pleiades open cluster, known as the Seven Sisters, is even younger than the Hyades is, around a hundred million years old. The photograph in **12.4** shows the central area of this cluster and the names of

12.4. The Seven Sisters of the Pleiades open star cluster. The Pleiades star cluster is about 400 light years away. NASA, ESA, AURA, Palomar Observatory.

the Seven Sisters. The stars identified as Pleione and Atlas in the image, although part of the central star group, are not part of the Seven Sisters. In Greek mythology, Pleione and Atlas were the parents of the seven sisters. To locate the Pleiades in the sky facing south, the Pleiades cluster is just above and to the right of Orion. Only six of the seven stars are clearly visible. Like the star Mizar in the handle of the Big Dipper, the seventh star is invisible except to those who have especially keen eyesight. The Pleiades cluster has fascinated observers since ancient times. The Greek Hesiod mentioned it as early as 1000 B.C.E., and references to the Pleiades are in Homer's *Odyssey* and in the Bible. It is referred to sometimes today by its Old European name of *Hen with Chicks*,

All stars are born in groups. Over time, a group of stars will evaporate, with individual stars going their own way depending on their mass and the gravity in different regions of the galaxy. It takes a long time for a group of stars to break up. Open clusters of stars, such as the Hyades and Pleiades, which have retained their groupings, are the home of some of the younger stars in the Milky Way Galaxy, and are therefore important astronomical star laboratories.

118

Galaxies

Across the vastness of the universe, the brightest stars are many billions of times less bright than the galaxies. Moreover, if the distances to the stars are unfathomable to the average observer, consider the galaxies. Their distances are truly beyond human comprehension. It would take more than 17,000 round trips to our closest star, Alpha Centauri, just to bridge a spiral galaxy like our own Milky Way Galaxy, estimated to be more than 150,000 light-years across.

It is estimated that there are in the universe more than 100 billion galaxies, some larger, some smaller than the Milky Way Galaxy, and each made up of billions, if not hundreds of billions, of stars.

Most Galaxies are generally flat, disc-like radial collections of stars and other cosmic matter. More than

100 times in diameter to thickness, a typical spiral galaxy is 50,000 light-years in diameter (5E+17 kilometers). Three typical spiral galaxies are in **12.5**. Their catalog numbers identify these galaxies. From the top: ***a*** Southern Pinwheel Galaxy M83; ***b*** Whirlpool Galaxy M51 and its companion galaxy NGC 5195; and ***c*** NGC 891. NGC 891 is in the Andromeda constellation, considered a twin of the Milky Way Galaxy.

While the study of individual stars provides clues to the composition and age of the known universe, the galaxies themselves give a large-scale structure to the cosmos. In their shapes they offer a comforting reference to Earth's experiences because many galaxies are remarkably regular in form, many appearing as flat spiral shapes whose tentacles reach out in patterns like dust from a spinning pin wheel.

The discovery of galaxies is a relatively recent occurrence dating to the early part of the 20th century. Consequently, many of them do not bear popular place names such as the ancients might have adopted, rather they bear the catalog numbers that modern astronomers use. The galaxies in **12.5**, for instance, have names like NGC 891, M83, and M51.

12.5. Three spiral galaxies. *a* The Southern Pinwheel Galaxy M83; *b* Whirlpool Galaxy M51, or NGC 5194, and its companion galaxy 5195; and *c* NGC 891. NASA, ESA, Hubble, Adam Block.

Since we cannot see our own galaxy from the outside, we must look at other galaxies to get an idea of what it looks like. The Milky Way Galaxy is of the spiral type, thought to be of similar shape and appearance to those shown in **12.5**. It looks a lot like NGC 891.

Meanwhile, Galaxy M31, alternately known as NGC 224, or the Great Andromeda Nebula, shown in **12.6**, is the nearest spiral galaxy. Like NGC 891, it, too, is very much like the Milky Way Galaxy, only considerably larger.

One galaxy name that has survived is Andromeda. The galaxy takes its popular name from the Andromeda

constellation named for an Ethiopian princess chained to a cliff for a sea monster to devour. The daughter of Cepheus King of Ethiopia and his queen Cassiopeia — also names of constellations — Andromeda has her constellation next to that of Perseus who, according to Ovid's Greek mythology, saved Andromeda's life and married her.

12.6. The Andromeda Galaxy, also known as M31, or NGC 224, is the nearest major galaxy to the Milky Way. NASA.

The ubiquitous presence of the Andromeda Galaxy in science fiction and space-age film comes most likely from its unique relationship to our galaxy. The largest of the nearby galaxies and at about 1.5 trillion kilometers in diameter, Andromeda has a structure very much like our own Milky Way Galaxy, with close similarities in both size and shape. Approximately one and one-half to two times as massive as the Milky Way, Andromeda is a spiral-shaped galaxy with a visible bright central nucleus.

It contains some 200 billion stars. Its rotation in galactic years is about like our galaxy: 200 million Earth years. Andromeda is the most-distant object visible with the unaided eye. Its distance variously calculates from 20 to 25 million trillion kilometers. Currently the agreed distance is 20.8 million trillion kilometers, or 2.2 million light-years. Only the central core of the galaxy emits light that reaches us because its outer parts are too dim. If we could see the entire disk, it would take up a space seven times the diameter of a full Moon. There is evidence that Andromeda, as it exists today, is actually the product of the collision of two galaxies. This has prompted astronomers to wonder if the Milky Way Galaxy may someday, billions of years from now, itself collide with another galaxy. We appear now to be moving at high speed toward Andromeda. Will Andromeda one day be a combination of not two but three galaxies?

119

Galaxy clusters

The Milky Way Galaxy is part of a cluster of galaxies called the Local Group, a collection of 18 to 20 galaxies moving together, like an open star cluster, through the universe. The largest galaxy in the Local Group is Andromeda. The two closest galaxies in the group in distance to the Milky Way are the Magellanic Clouds. In (**12.7**), the Small Magellanic Cloud is on the left; the Large Magellanic Cloud is on the right. The Local Group is part of a larger group of about a thousand galaxies in the Virgo super cluster.

The Virgo galaxy cluster is itself just one of many super cluster groupings. The Virgo cluster and many other large super clusters of galaxies like it throughout the universe appear to be moving at great speeds toward a gigantic unseen mass of energy dubbed The Great Attractor. To the naked eye, galaxies look like stars, because at such great distances their luminosities are mere points of light in the sky. While we cannot see individual stars in other galaxies, galaxies themselves are visible with the unaided eye. The Andromeda Galaxy, for example, is 2.2 million light-years distant, yet it shines clearly to the unaided eye. Beyond the range of the Local Group, however, even giant galaxies are invisible without a telescope.

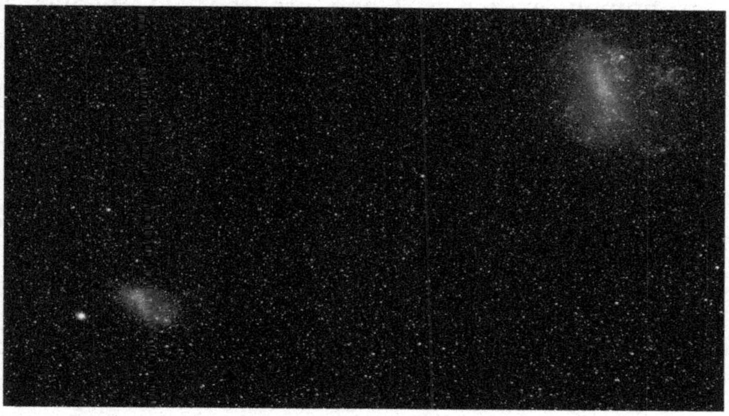

120

12.7. The Large and Small Magellanic Clouds. European Southern Observatory.

The story of modern astronomy is the story of deep space exploration. Science has reached beyond the stars, beyond the nearby galaxies, to the very edge of the universe and perhaps to the beginning of time.

Earth, situated about two-thirds of the distance out from the center of the Milky Way Galaxy, establishes the place from where we view the rest of the universe. As we reach out across the vastness of space, our vision encounters the great objects of the cosmos that stand as timeless sentinels between human mortality and infinity.

If the ancient astronomers returned today, they would find that the heavens appear about the same as in their day in early history. However, the understanding of the universe has radically changed from its first interpretations. Earth's location in the universe today has it orbiting an unremarkable star situated in an unremarkable galaxy, very different from what early astronomers thought. Ptolemy convinced the world that the Sun revolved around Earth. He would be shocked to learn that not only is Earth not at the center of the universe, as he believed it to be, but there is nothing particularly special about Earth's location at all, or that of the Sun that sustains it.

Modern understanding of the universe has provoked remarkable changes in the perception and interpretation of human existence. Gone are the old theological doctrines that once said humans enjoyed a special relationship with the universe. Perhaps time may reveal that humans actually are a chosen species, but if so, it will not be because of their unique physical placement in the cosmos. Looking now at the universe, human beings do not appear to be at the center of anything.

Newborn Star Inside the Milky Way Galaxy.
Hubble Space Telescope.
NASA Goddard.

Stellar Nursery NGC 3324 in the Carina Nebula. Cosmic cliffs of ionized gas and dust seven light years high. James Webb Space Telescope. NASA, ESA, CSA, and STScI.

Galaxy NGC 1961, located about 180 million light-years from Earth in the constellation Camelopardalis.
Hubble Space Telescope.
NASA, ESA, J. Dalcanton, et al.

XIV

BIRTH OF A STAR

121

Immediately after the Big Bang, the temperature of the universe was approximately a trillion degrees centigrade. Cooling began immediately. Within a fraction of a millisecond, the properties of the universe came into existence. Moreover, within seconds of its explosive beginning, the basic atomic particles of the universe were active. The temperature dropped to a billion degrees in the first 100 seconds, and in minutes, the universe had cooled to 100 million degrees. The elements began to form as energy coalesced into matter that continued to cool and build mass. Gravity assumed the fundamental role of holding together the developing bodies.

122

In the first million years or so after the Big Bang the expanding universe remained densely packed, as much as 1E+93 grams per cubic centimeter, or roughly equivalent to compressing the total mass of 1E+65 Earths (one followed by 65 zeros) into a space the size of a pea. Photons, ordinarily the transporters of light through space, were quickly absorbed; the universe remained dark and opaque. As it grew, the universe cooled to about 3,000 degrees. Elements began to form, and the voids of space suddenly appeared. The universe became transparent. Photons became free, released to move from one place to another. Today, after billions of years in an ever-enlarging universe, photons are free to travel great distances, carrying with them information about the earliest moments of time.

123

A star begins its life as a glitch in a sea of cosmic gas distributed uniformly throughout the universe. Because of this glitch in uniformity, matter begins to collect, developing as a gravity center, which occurs more or less anytime one part of the cosmos becomes more massive. Gradually enough gas is collected that the growing gravity field begins to pull harder and to shrink and compress the gases into a denser and denser body, on its way to becoming a star. As compression continues, the internal temperature of the collecting mass starts to rise. At some point, compression is sufficient to ignite nuclear fusion. When this happens, a star is born. Nuclear fusion transforms the original cosmic gases — made up mostly of hydrogen atoms — into helium and other elements that will sustain the star for the rest of its existence.

As nuclear fusion builds within a new star, some energy begins to escape the star's gravitational field. This is what we see as star light and detect as other forms of radiation. When hydrogen converts into helium, the original hydrogen atoms have all the energy needed to become helium, with a little left over. This remnant accounts for the radiant energy of the stars. A star will release as much energy in one second as all the world's power plants in a year.

The compression of a star will continue to raise its internal temperature, until its level of nuclear reaction is equal to the amount of radiation lost. When the star reaches this state of equilibrium, it is at *zero age*. It completed its proto star stage to become a fully developed star. The whole process in the birth of a star may take 10 million years or more, in a star life that could last more than 10 billion years.

One of the many places in the universe where new stars are being born is the Great Orion Nebula pictured in **13.1**. The Orion Nebula is a mass of hot cosmic gas

13.1. The Great Orion Nebula in the constellation of Orion. NASA, ESA, Hubble Space Telescope.

and one of the most active regions of new star formation. Located in the constellation of Orion, the Great Nebula is visible from the northern hemisphere in the southern winter sky. It is the fuzzy star in the middle of the sword of Orion, just south of the three main stars of Orion's belt. Considered the finest example of a diffused nebula in the sky, and the most famous of all astronomical nebulae, the Great Orion Nebula is a virtual star nursery. Astronomers have even detected the formation of new planetary systems there. It is estimated that as many as 10,000 stars similar to our Sun may be formed in the Orion Great Nebula over the next several billion years.

Astronomers have already detected more than 2,000 new stars near its center, each one less than a million years old.

One of the reasons we know so much about Orion is because of its relatively close proximity to Earth. About 1,500 light-years distant, Orion is located in the same vicinity of our galaxy as the Sun, although positioned in the next spiral arm over. The view shown in **13.1** represents only about 2.5 light-years of the inner portion of this 30-light-year diameter nebula.

124

The zero age main sequence

In the year 1910, the Danish astronomer, Ejnar Hertzsprung, unveiled a study that would become a benchmark of modern astronomy. It correlated the brightness of stars with their respective temperatures and concluded that the large majority of stars conform to predictable temperature and brightness relationships. At the same time that Hertzsprung made his discovery, an American astronomer, Henry Norris Russell, independently completed a similar project at Princeton. The results of the work of Hertzsprung and Russell produced a graph. In it, the luminosities of stars appear plotted as a function of their temperatures. A version of the graph, known as the Hertzsprung-Russell, or H-R, curve is in **13.2.**

Newborn stars receive a place on the H-R curve according to their color. Knowing the color or spectral type of a star makes it possible from the H-R diagram to estimate its mass. Stars that are more massive are bluish in color and less massive ones are red. The blue star Regulus, for example, has a mass about 20 times that of the Sun. Barnard's Star, on the other hand, is a red star of about one-third the mass of the Sun. These and several other well-known stars, including the Sun, are in **13.2** plotted relative to the H-R curve. Almost all stars conform to this curve. Those that do not are usually in

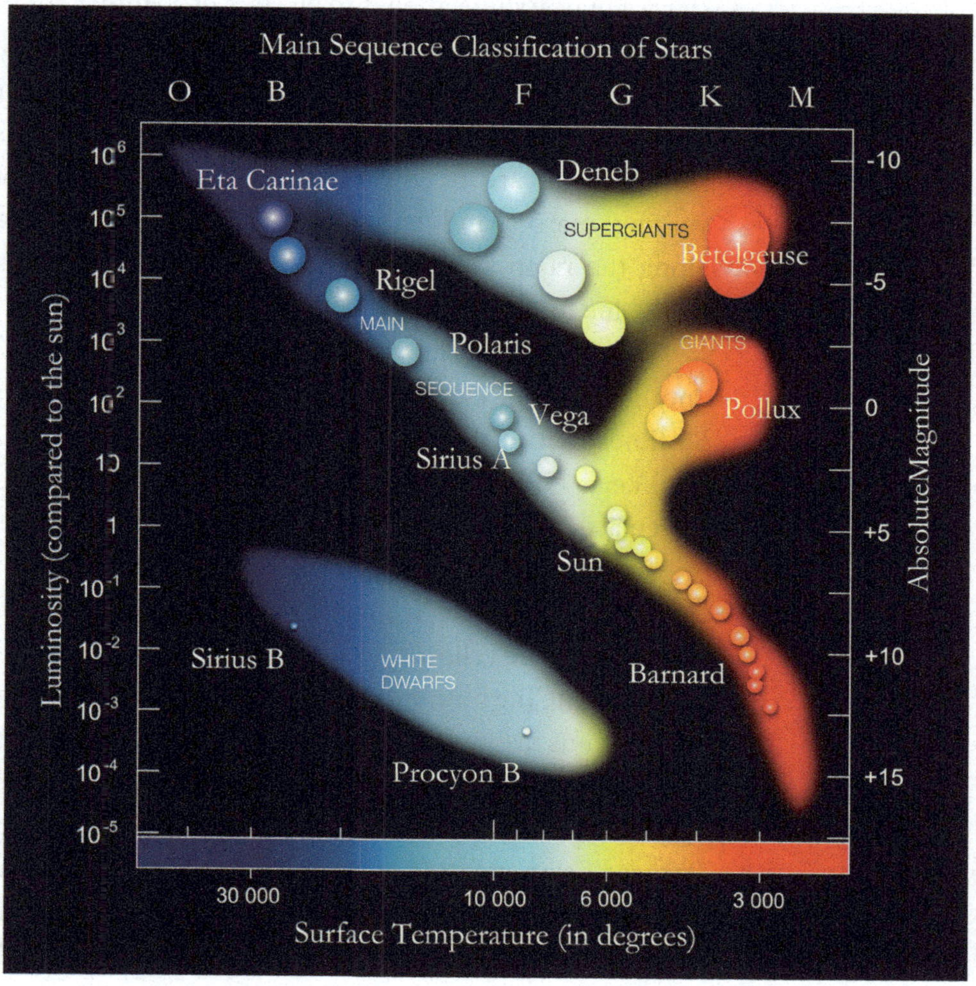

13.2. Hertzsprung-Russell zero-age main sequence diagram. European Southern Observatory.

advanced stages of their evolution, some nearing extinction. Today the H-R curve is central to predictions about the nature of the universe, because once we know the mass of a star, many things about its behavior and composition is predictable.

125

The significance of the main sequence H-R curve lies in the discovery that between the birth and death of a star, 90 percent of all stars will fit somewhere along the main sequence curve. Most stars are very much alike in their evolution, and the color of a star is not only a reliable predictor of its mass, but also of its temperature, composition, and luminosity. What we learn about one star we may assume applies to all stars of a similar color. The main sequence graph shows that blue, more luminous stars have higher temperatures and greater masses; red stars have cooler temperatures, are less massive, and are of lower brightness. Other stars of varying masses and temperatures appear distributed along the main sequence H-R curve in colors varying from blue to red across the color spectrum.

Vega, the first star photographed back in 1850, is another example of a blue star. With a luminosity that is 58 times that of the Sun, Vega is three times the Sun's size. Vega, Polaris the North Star, which Vega is destined to replace in the future, and Sirius, to name just three, are all very hot stars that are shaded toward blue. Our own Sun, meanwhile, is more in the middle of the color spectrum in the yellow to yellow-green region.

Opposite the blue end of the curve, many of the stars closest to us are red. Known as flair stars, or red dwarf stars, these main sequence red stars are much smaller than the Sun and too faint to see with the unaided eye. Barnard's star, mentioned above, is a red dwarf star. Red dwarf stars are different from red giant stars in that, unlike red dwarfs, they are not main sequence stars at all but are instead very bright, evolving stars that exist outside the normal H-R main sequence curve.

126

Invisible color

When we view stars with the unaided eye, they appear to be of different magnitudes but not necessarily of different colors. Their colors are generally not easily detectable because at night we see with the *rod* system of the eye, the part of the eye that sees brightness but not hue. Hue is the aspect of seeing that distinguishes one color from another, seen by the other part of the visual system called *cones*. Below a certain threshold of light, the cones do not work, and we do not see hues. We see only the brightness detected by the rods.

At the same time, the optical radiation of a high temperature star appears truncated away from its natural blue state by Earth's atmosphere. Absorption of colors by the atmosphere, particularly in the blue range, seriously limits the color appearances of stars. This filtering action also affects the color of the Sun. The blue component of sunlight is scattered by Earth's atmosphere, so the solar disk appears yellowish, or whitish in color, just as it appears redder in the morning and evening because its blue light must pass through greater amounts of the atmosphere because of the Sun's angle. Figure **13.3** illustrates the colorful nature of the universe in a picture of a small piece of the universe taken by the Hubble Space Telescope Deep Field camera.

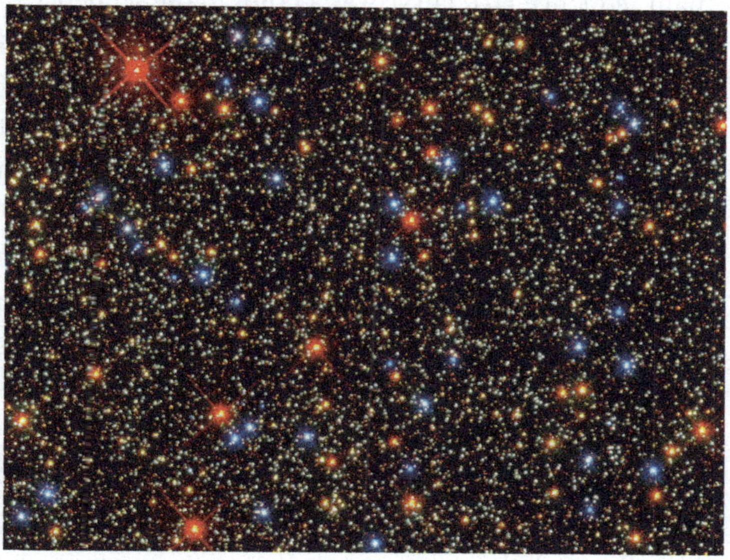

13.3. Colorful stars inside the globular star cluster Omega Centauri in a panoramic view of 100,000 stars taken by the Hubble Space Telescope. NASA.

The Hubble's color photograph reveals colors that are not normally detectable visually. Photo sensors record light at exposure levels well below the threshold of color vision. Atmospheric absorption, coupled with the adaptive qualities of the eye, makes it virtually impossible to see, as a camera does, the myriad colored stars that dot the universe. Consequently, stars appear white to the naked eye, with only the faintest tinges of blue or red. This is the normal response. The heightened visual sensitivity of an artist, like Van Gogh, may be an exception. Perhaps it was his imagination that caused him to write about the "blue whiteness of the Milky Way" and to see in the blue depth the stars "sparkling–greenish, yellow, white, rose–brighter, flashing more like jewels...opals you might call them; emeralds, lapis, rubies, sapphires." It is enlightening to believe that Van Gogh could have such an uncommon ability to see what most cannot, like those who claim to feel the motion of Earth moving through the universe. Sometimes our enlightenment comes from what only a few of us dare to imagine.

127

Stellar mass and the life of a star

At first, scientists thought that each point on the main sequence H-R curve was a point in the sequence of the life of a star, that is, a star began its life very hot and gradually cooled as it consumed its fuel, until it extinguished and died. In effect, analysists believed that a star moved sequentially down the H-R curve as a function of age.

We now understand that each point on the main sequence curve correlates instead with a star of a given mass. A star born at a particular temperature maintains that temperature throughout the duration of its life. Stars do not change places on the main sequence curve. Because a star ignites at a certain temperature where it remains, the main sequence graph of Hertzsprung and Russell we call the *zero age* main sequence graph.

128

The mass of a star depends on how much matter collects around its gravity center before nuclear fusion begins. How long a star lives is dependent on its mass, as is the nature of its death.

The mass of the Sun calculates to be 1.9891E+33, or 333,000 times the mass of Earth. Astronomy adopts the Sun's mass as a convenient reference for designating the masses of other cosmic bodies; for this reason, mass of cosmic bodies is in solar masses. The mass of Earth, for example, is 1/333,000 solar masses. However, just because the Sun is more massive and larger than Earth with a mean diameter of 1.4 million kilometers, or 110 times the diameter of Earth, it does not mean that the size of every cosmic object necessarily is proportional to its mass. Mass should not be confused with size. More than one million Earths will fit within the Sun, but the Sun is not a million times as massive as Earth. It is even possible for a more massive object to be smaller than a less massive one; that is, an object of much greater mass than Earth can be much smaller than Earth.

This is a somewhat confusing concept because all the objects in our solar system conform to the expected size-mass relationship, with larger objects having more mass than smaller ones. The Sun and larger planets have more mass than the smaller planets, although not necessarily proportional to their relative sizes. Yet the idea of smaller, more massive objects need not be confusing. On Earth a cubic centimeter of lead has substantially more mass than a much larger 10,000 cubic centimeters of helium, because the atoms that make up each of these elements — lead and helium — have radically different spatial distributions. Atoms in helium are less complex and much further apart in their gaseous state than the atoms that make up solid lead.

129

The density of mass

The distribution of mass in a volume of space is density. The more tightly packed the mass, the greater is its density. This in turn affects gravitational force. Density is critical to the function of the universe because the greater the mass the stronger is the potential gravitational field. Moreover, if a gravity field is contained in a relatively compact, dense area, its force amplifies.

We measure density in grams per cubic centimeter; with the standard being, that one cubic centimeter of water is equal to one gram. Just as the temperature scale calibrates to the freezing point of water, water also serves as a unity reference for density. This allows one to compare the relative densities of other materials to water, which is a well-known, common substance.

Earth has a density of 5.5 grams per cubic centimeter, which is roughly the average density of the materials composing it; viz., rock at 2.5 to 3.2 grams per cubic centimeter, and metals at an average of about eight grams per cubic centimeter, although lead weighs in at 10 grams. Water as noted has a density of one gram per cubic centimeter.

The Sun and the larger planets: Jupiter, Saturn, Uranus, and Neptune all have densities much less than that of Earth and the inner planets. Uranus, for instance, has a density of 1.6 grams per cubic centimeter, as does Neptune. Saturn, the least dense planet, weighs in at 0.69 grams per cubic centimeter. Saturn is not even as dense as water, which means that large portions of it are still in a gaseous state. The Sun and Jupiter, meanwhile, have about the same density, 1.4 grams per cubic centimeter for the Sun, 1.33 for Jupiter. The mass and density of Jupiter are of special interest, because Jupiter came very close to being a star, close at least on an astronomical scale.

130

Almost a star

Jupiter needed only 75 times more mass to ignite as a star and commence the nuclear fusion process. If just 7 percent of the Sun's mass had wound up in Jupiter instead of the Sun, we would now have two Suns instead of one. Of course, this would completely change the delicate balance of the solar system, as we know it, to the extent that an earth-like planet might or might not have formed.

As it is, Jupiter radiates twice as much heat as it gets from the Sun, with temperatures varying from thousands of degrees in the interior of the planet to below freezing in its upper atmosphere. Still, Jupiter did not ignite to become a self-luminous body, and what we see in the sky as the giant planet is sunlight reflected from the Jupiter atmosphere. A telescopic image of Jupiter and two of its famous Moons are in **13.4**. The Moons Io and Ganymeade show up silhouetted against the blackness of

13-4. Planet Jupiter and two of its Galilean moons, Io and Ganymede. Auroras in bright orange extend to high altitudes above both the northern and southern poles. Jupiter dominates the black background of space. The Great Red Spot, a famous storm so big it could swallow Earth, appears white in this view, mapped in visible color by the Webb Space Telescope and shown below in its familiar colors. NASA, ESA, CSA, Jupiter ERS Team. Composite photo after Judy Schmidt and Damian Peach.

space. The Moons of Jupiter have been a curiosity ever since Galileo first discovered them with his newly invented telescope in 1610. Jupiter is so large compared to other planets in the solar system that some of its Moons are larger than the planets Mercury and Pluto.

131

Brown dwarfs

Jupiter is at one end of a class of stellar objects called brown dwarf stars. Brown dwarfs form out of hydrogen gas just like stars. However, unlike stars, brown dwarfs do not contain enough mass to ignite their nuclear engines. So they exist somewhere between stars like the

Sun and planets like Jupiter. After looking for years for these elusive, theoretical objects in space, astronomers finally located one in 1994. Since then discovery of hundreds occurred, particularly in new star regions like those of the Great Orion Nebula. The brown dwarf discovered in 1994 is calculated to be as much as 50 times as massive as Jupiter but about the same diameter as Jupiter. To give some idea of how this compares with the range of masses found in the solar system, the Sun, which is 1,000 times as massive as Jupiter, has more than 99 percent of the mass of the entire solar system. Less than 1 percent of the system's mass is in the planets. Of the 1 percent, Jupiter has more than two-thirds of the total mass of the planetary system, and 318 times more mass than Earth, with a diameter eleven times that of our planet. Fourteen hundred Earths would fit comfortably inside Jupiter's volume. Yet for all its impressive size relative to Earth, Jupiter has only $1/1000^{th}$ the mass of the Sun.

Even had Jupiter become a star, its independent existence, as a companion to the Sun is far from certain. Stars and galaxies, too, for that matter, draw together and sometimes collide because of their shifting masses and occasional chaotic trajectories to reform as new cosmic bodies in a continuously changing universe. A more massive Jupiter, either as a brown dwarf or a full-fledged star could easily have succumbed to such a fate because of its relatively close proximity to the Sun. Brown dwarf stars, even within a common region of star formation, are much more widely dispersed than planets like Jupiter.

132

The solar system is stable. The Sun is in the middle of its stellar life, and astronomers predict no great cosmic changes that would drastically alter our system.

Nevertheless, in other parts of the Milky Way Galaxy and throughout the universe this is not always the case; matter is in a constant state of dynamic change. Stars are born, evolve, and die, while entire galaxies collide in spectacular events that redistribute mass and alter the patterns of density in the universe. Figure **13.5** pictures one such event where two galaxies approach each other, attracted to each other by their gravitational fields.

Intersecting galaxies like these give off brilliant displays of light. However, unlike exploding stars, colliding galaxies avoid characterization as violent changes. Stars in colliding galaxies are located trillions of miles apart and seldom collide with each other. Instead, colliding galaxies pass through each other, changing over time, as energy patterns disrupt and rearrange matter. The intersecting regions of mass, particularly in the interior regions of colliding bodies, produce some of the most active star producing regions in the universe. The intensity and brightness of these star-forming regions give to them the name *starbursts*.

13.5. Colliding galaxies in the constellation Hercules, about 230 million light-years away. European Space Agency, Hubble Space Telescope, NASA.

133

An open universe

Theories of the evolution of the universe rest on the packing and unpacking of mass. When enough of some kinds of elements pack into a critical density, they ignite into nuclear fusion and become stars. Most of the known mass of the universe is in the stars. Therefore, if we divide estimated mass of all the stars in the universe by the volume of the known universe, a hypothetical average density of the cosmos obtains. In this way, the density of the universe projects to be approximately 1E-31 grams per cubic centimeter, or roughly equivalent to spreading the mass of a grain of sand through a volume the size of 10 thousand Earths.

134

The density of the universe is important, because below a certain density we consider the universe *open,* while above a certain level we deem it *closed*. These two contrasting theories of the state of the universe, open versus closed, simply mean that there may or may not be sufficient mass in the whole of the universe to create the gravitational field, or fields, necessary to reverse the outward expansion of the universe.

The open and closed theories of the universe are not the only theories about its state of existence. A third theory is *steady state.* It gained popularity in the late 1940s. Steady state theory assumes that the universe is perpetually in balance, with new matter created as needed to fill the voids of space. Modern discoveries, however, appear to support the Big Bang theory for a common origin of all matter, and have generally discredited the steady state theory. Discoveries by the Hubble Space Telescope have confirmed that the universe is not only expanding, it appears to be accelerating, thus further complicating efforts to unravel the secret of its origin.

An open universe will expand indefinitely, while a closed universe is destined at some point to reverse itself and begin contraction in 40-50 billion years. Discovery of the universe so far declares that mass is insufficient to stop the expansion of the universe, and the universe is therefore open. The most optimistic estimates can account for only less than 40 percent of the density of mass required for bringing the universe back to closure. Research continues toward locating additional mass that would be sufficient to reverse present theory, thus allowing closure of the universe to begin to contract under its own gravitational force toward what modern astronomers have dubbed the *big crunch*, a phrase intended to complement the theory of the Big Bang.

One place astronomy looks for the needed missing mass is in so-called *dark matter*, matter undetected by conventional means. Some models of the universe predict as much as 95 percent of the existing mass to be in yet undetected dark matter. What dark matter is, or could be, in amounts large enough to close the universe remains largely a mystery.

Center of the Lagoon Nebula in the Constellation Sagittarius. NASA/ESA Hubble Space Telescoe. NASA, ESA, J. Trauger (Jet Propulson Laboratory).

XV

STELLAR CLASSIFICATION

135

Main sequence stars — those mature stars still in the prime of their lives — may vary in temperature from about 2,000 degrees Kelvin to a theoretical 60,000 degrees; no one has yet discovered stars of the highest temperatures. The hotter the star, the rarer it is. Most star temperatures fall in a range from about 2,000 to around 20,000 degrees Kelvin. The Sun, which is toward the cooler end of the spectrum, has a temperature of 5,800 Kelvin.

Kelvin is the temperature scale used to describe star temperature. Kelvin correlates to star color. The color temperature of a main sequence star when applied to the main sequence H-R graph produces estimates of the luminosity, brightness, and absolute magnitude of the star. This, in turn, then yields an estimation of stellar distance for that star.

136

Temperature scales

When specifying temperature, it is necessary to know what temperature scale applies. There are three commonly used scales: Fahrenheit, Celsius, and Kelvin.

Galileo made the first thermometer in 1593. It was composed of a bulb of glass with a long stem submerged in water. As air in the bulb cooled, some of the water rose, drawn into the stem. The amount of water in the stem fluctuated in level according to the air temperature.

Unfortunately, the water level also changed with fluctuations in air pressure. Scientists in the 17th century partially eliminated this shortcoming by replacing water with alcohol in a sealed glass tube. Alcohol is a material still used in inexpensive thermometers. Today several heat-sensitive materials are in use.

In the 18th century, circa 1714, the German physicist Gabriel Daniel Fahrenheit introduced mercury as the heat recording substance. The introduction of mercury also saw the adoption of the Fahrenheit temperature scale, the oldest of the scales still in use.

In 1742, the Swedish astronomer Anders Celsius adjusted the Fahrenheit scale. This ultimately came to be the Celsius, or centigrade, scale. The basis of the Celsius scale is the metric system, hence its alternate name of centigrade. (While practically interchangeable, Celsius and centigrade actually vary in scale by about 0.1 percent of a degree.) The centigrade, or Celsius, scale provides fixed points of zero degrees and 100 degrees for the freezing and boiling points of water respectively. On the other hand, in Fahrenheit temperature, water friezes at 32 degrees and boils at 212 degrees.

Meanwhile, the Kelvin color temperature scale is yet another way to measure temperature. William Thomson developed the scale in 1848. An Irish-born British mathematician, physicist, avid inventor, and naturalist, Thomson was knighted Baron Kelvin and known thereafter to the scientific community as Lord Kelvin.

William Thomson — Lord Kelvin — was educated at home by his mathematician father. He entered the University of Glasgow at the age of 10. Although he did not receive a degree, he returned to Glasgow in 1841 to become professor of natural philosophy, a job he held for 53 years, moving up to chancellor of the university in 1907. Thomson held more than 50 patents and was profoundly interested in practical science, especially in the area of electricity. The British Crown knighted him because he oversaw the successful laying of the Atlantic cable in 1866 after an earlier failed attempt when others ignored his advice. Aside from the temperature scale that bears his name, he was the physicist who set the foundation for the second law of thermodynamics.

The Kelvin scale is the preferred scale in astrophysics because unlike the other two scales — Fahrenheit and Celsius — Kelvin is calibrated to measure absolute zero, which is set at minus 273 degrees Celsius; that is, 273 degrees below the freezing point of water, or minus 459.4 degrees Fahrenheit.

The scales of Kelvin, Celsius, and Fahrenheit provide different calibrations of temperature. The point at which water friezes, for example, is zero degrees Celsius, 32 degrees Fahrenheit, and 273 Kelvin. Celsius temperature converts easily to Kelvin by the addition of 273. The conversion of Fahrenheit to Kelvin is more complicated. Figure **14.1** shows how these scales are related. The diagram is for a 100-degree range on the Celsius scale. This range is typical of surface temperatures found on

14.1. Temperature equivalences for Fahrenheit, Celsius, and Kelvin temperature scales.

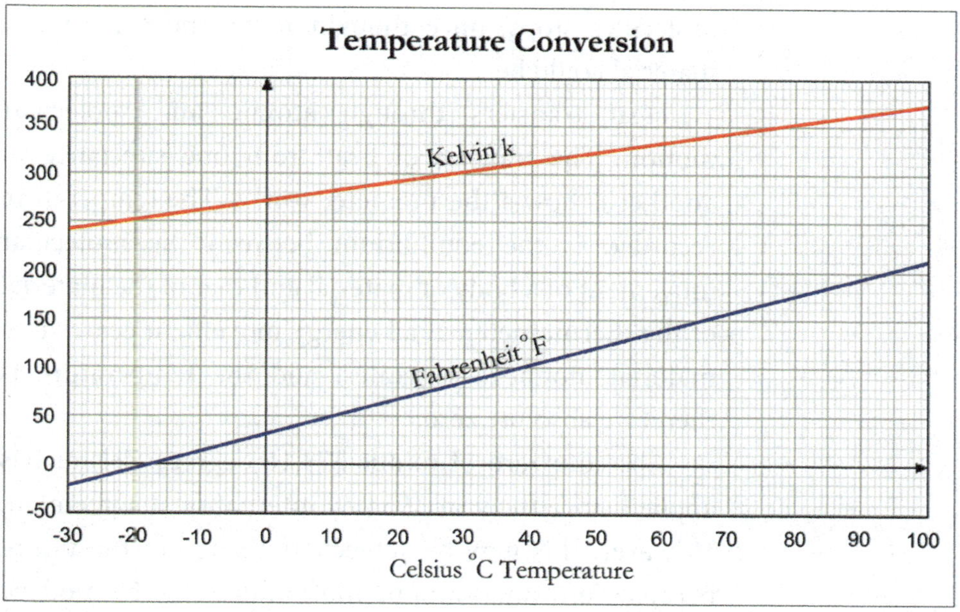

Earth. To convert from Celsius to Fahrenheit or Kelvin temperatures, read from the bottom Celsius scale vertically to intersect the diagonal lines, then left to find the equivalent temperature in either Fahrenheit or

Kelvin. Reverse the process for Celsius in Kelvin or Fahrenheit temperatures.

Indicated by the letter *K*, Kelvin temperature is always the scientifically preferred scale of measurement in celestial temperature. Astronomers prefer the Kelvin scale to the Celsius and Fahrenheit scales. In many cases, when measuring the extreme temperatures of stars, a difference of 273 degrees from the Celsius scale to Kelvin is not significant. However, sometimes a more exact reference is helpful. For example, it has been determined that the temperature of the universe is -270 degrees Celsius. However, instead of saying the universe has a background temperature of -270 degrees Celsius, using Kelvin temperature one can say its temperature is 3K, which means the universe is only three degrees above absolute zero.

Absolute zero temperature is the point at which all-available energy from the random motion of atoms and molecules stops, once thought to be the coldest any material could be.

Heat transfers through atomic and molecular motion. The transfer of heat from one substance to another is how a thermometer works. The temperature recorded is the equilibrium between the molecular movement of a body of material and that of the material of the thermometer. If all energy movement ceases in a material, no heat generates, and the temperature is therefore absolute zero.

Absolute zero of minus 273.16 degrees Celsius has applied since the year 1954 by international agreement. However, this may be subject to change in the future because zero-point atomic motion detected by modern quantum mechanics exists in matter at temperatures below Kelvin absolute zero.

Cosmic temperature relates to background radiation in space. Remnants of the original radiation from the Big

Bang still exist, traveling through space, and continuing to cool. Scientists working independently predicted cosmic background radiation as early as 1948. Bell Laboratory scientists confirmed its existence in 1963 while trying to correct what they thought was radio noise, but which turned out to be of cosmic origin. Background radiation defines the standard of rest of the universe by which *absolute velocities* of cosmic bodies take measure. Absolute velocity is velocity relative to the primeval background radiation of the universe, which defines the standard of rest of the universe.

Background radiation appears evenly distributed in all directions, having a uniform temperature of about 3K, although it is slightly warmer in one direction and slightly cooler in the opposite direction.

The general uniformity of space as seen in its temperature presents a dilemma in its apparent contradiction of the Big Bang theory. While the existence of background radiation is consistent with the Big Bang, its uniformity is not. For stars and galaxies to form there must be a certain lumpiness inside the uniform expansion of the universe. Clumping of mass will result only where there are variances in background radiation. Big Bang theorists accommodate the problem by asserting that the background is smooth, but not entirely smooth. Given the vastness of the universe, this is probably a reasonable explanation.

137

Color temperature and magnitude

We estimate the temperature of a star by observing its color; color and temperature relate to luminosity; and luminosity is the basis of brightness. Brightness, of course, directly correlates with the magnitude scale. By applying the inverse square law to the apparent magnitude of a star, it is possible to arrive at its absolute magnitude as predicted by the main sequence H-R curve. This method obtains a good approximation of the distance to a star. The use of color-temperature analysis to predict cosmic distances is *spectroscopic parallax.*

Today, techniques such as spectroscopy, originally developed solely for the analysis of light, include advanced processes to accommodate other parts of the invisible electromagnetic spectrum. For centuries, the spectroscopic study of the universe relied entirely on

14.2. Spectrum of a lightning flash. ***Britannica*, William S. Bickel, University of Arizona.**

light and color. In the remarkable photograph in **14.2**, the atmosphere disperses the brilliance of a lightning flash into a spectrum of blue, green, and red, colors known as the primary colors of light. It was observations like the rainbow, and the famous prism experiment of Isaac Newton that eventually led scientists to the conclusion that the different colors in light represent different energy levels, which in turn led to color-temperature analysis and the sciences of color and astrophysics. Consequently, over time, more knowledge accumulated about the universe through color than by any other means. No doubt, it was observations of phenomena such as lightning and the natural rainbow spectrum of the Sun that led Isaac Newton to his famous experiment to show ultimately, how all light is composed of different colors.

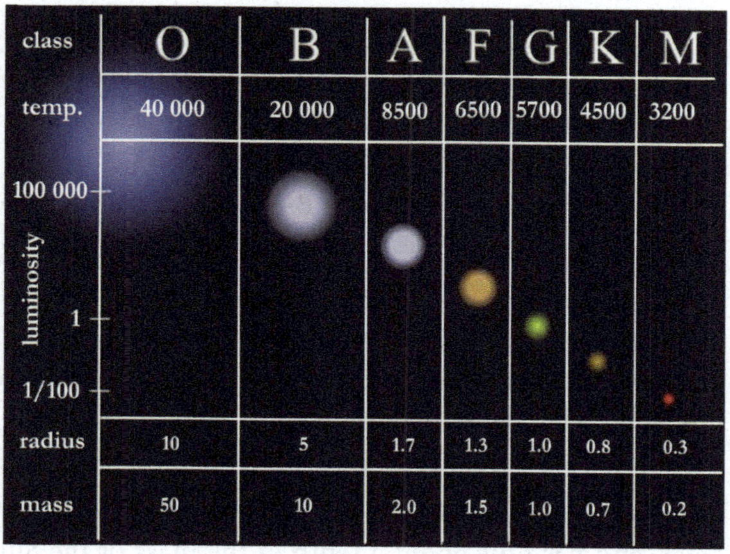

14 3. Table of star classification.

Stars classify by color temperature. Stars in each temperature and magnitude class receive unique color category designations. Table **14.3** shows the relationship of star identification characteristics for main sequence stars. The table gives approximate temperature and color relationships for each class of star, along with size and mass relative to Earth under the letter codes adopted by astronomers for each classification level. Shown for comparison also is how much more or less luminous a particular class of star is relative to the absolute magnitude of the Sun. Stars with smaller absolute magnitudes are more massive hence more luminous than the Sun.

In the star-classification letter system, the Sun is a G-class star; Polaris the Pole, or North Star, is an F-type star, while Barnard's star is an example of a lower energy M-class star emitting its energy in the red part of the color spectrum. The highest energy O-class star emits most of its radiation in the invisible ultraviolet spectrum. O-class temperatures are so high that much of their energy is beyond the visible range of the human eye. The hottest stars are therefore sometimes less visible than stars in other classes. That is to say, the hottest stars give off more energy, but less of it is visible. With a mass of up to 50 times that of the Sun, or more, an O-class star if fully visible would be at least 10,000 times as luminous as the Sun, perhaps as much as 100,000 times.

Just because O- and some B-class stars are invisible at their peak radiation emission levels does not mean they stay undetected. Although optical devices cannot provide an accurate measurement of such higher temperature stars, instruments are available to measure the higher energy radiation.

In contrast to O-type and B-type high-energy stars, Barnard's star, an M-class star, is a red main sequence star, and has only about a tenth of the mass of the Sun and about 1/100th of the Sun's light. It, too, is difficult to see due to its low luminosity, and we know it mostly by its invisible radiation.

138

Black body temperatures

One method to determine the temperature of a star by looking at its color is through the application of a physical phenomenon of matter known as blackbody radiation. As an object grows hotter, it changes color in very predictable ways. We observe this phenomenon in the light emitted by a piece of heated metal (**14.4**). At first the metal glows with a dull red color. As it grows hotter the color changes from red to orange, then yellow, and finally white as the metal begins to melt, or disintegrate. The first color seen is a faint red, indicative of low energy emission of light. As the temperature of the iron increases, color shifts toward the higher energy,

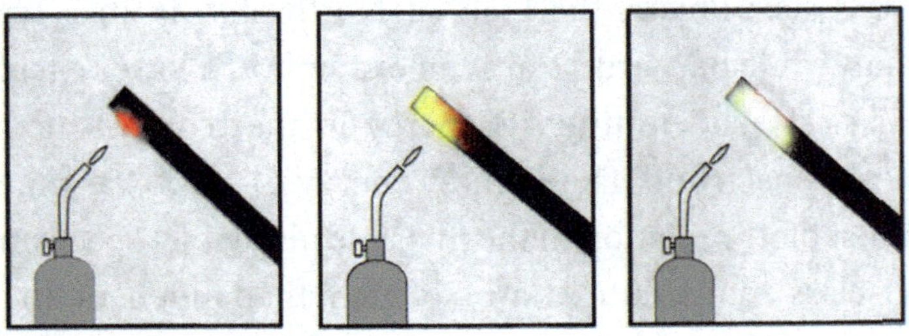

14.4. Color temperature changes in a piece of heated iron.

yellow portion of the visible spectrum. Eventually, if sufficiently heated, the iron becomes white hot. The peak of the energy emitted is in the middle of the visual spectrum, balanced by lesser amounts of red and blue energy to produce the resultant neutral white color. Like the Sun, the heated iron has its peak in the green spectrum band but appears white because black bodies radiate at all colors at the same time, some colors with greater intensity. This maximum color intensity is the radiation peak. In the example of the piece of iron, if more heat could be added without melting it, the iron would glow blue-white, a still hotter temperature indicative of higher energy. Were it possible to preserve metal at high temperatures, a heated body would glow blue, until at its highest temperature it would have its peak radiation in the ultraviolet range. Metal, of course, disintegrates by heat long before it reaches these high temperatures. Stars, on the other hand, can produce and withstand extreme temperatures through their internal nuclear fueling process.

Stars are natural blackbody radiation emitters. While there is no such thing as an absolutely perfect black body, a star comes very close to being such a body, close enough that its compliance with blackbody behavior theory is accepted, viz., blue stars are hotter black bodies; red ones are cooler. A blue-white star is much hotter than a reddish one. As a blackbody radiator, our Sun is somewhere in the middle of the color spectrum.

139

The technique for the evaluation of the color temperature of light comes from the groundbreaking discoveries of the German physicist Max Planck. Building on the work of his Prussian-born colleague, Wilhelm Wein, and others, Planck showed how the frequency of oscillation in a light wave relates to the relative energy of the radiation.

Planck, shown in **14.5**, was an individual of exceptional talent. He was an excellent pianist blessed with perfect pitch. It was only after much consideration that he opted for a career in physics rather than music. He received his doctoral degree in physics from the University of Munch at the young age of 21. In 1885, he returned to his home town of Kiel to take a job at the University of Kiel. Later, he moved to the University of Berlin where he did his groundbreaking research. In 1900, he derived the formula to explain the behavior of blackbody radiation for which he won the Nobel Prize for Physics in 1918. One of the striking revelations about Planck is that he deduced his formula for blackbody radiation in less than two months, upsetting centuries of classical physics and ushering in the age of quantum theory.

14.5. Portrait of Max Planck. Unknown photographer, c. 1933.

A member of a large, devoted, and well-educated family, tragedy marked Planck's personal life. He lost a number of his children at early ages. The Nazis executed another son when accused of a failed attempt to assassinate Hitler in 1944. Rescued by the American Allies, Planck had nonetheless lost the will to live and died in 1947 at the age of 89.

Planck's conclusion that heated matter emits radiation of discretely different energy levels, hence different temperatures, resulted in the blackbody, or Planck, radiation curve, and eventually in the development of the theory of 20th century quantum physics. Known as the father of quantum physics, Planck opened the way for modern understanding of the universe.

Energy is not a continuous stream of radiation but rather a discontinuous flow of discrete, indivisible units of propagation. In light, unit propagation of energy occurs in the behavior of photons and electrons. Electrons are negatively charged particles that orbit the nucleus of an atom. Electrons oscillate on and off in very rapid sequence. The frequency of the vibration of the electron precisely matches to its orbit. That is to say, in order for the orbit of the electron to close on itself without interruption of the oscillation, the orbit and frequency of oscillation must be interdependent.

Light energy, meanwhile, travels in the form of another type of vibrating energy called a photon. Electrons absorb photons. When the frequency of oscillation of a photon matches that of the electron, the electron will absorb the energy of the photon, that is, the photon will add one or more vibratory links or chunks to the orbit of the electron. The expanded energy level takes the electron into a higher orbit around the atom's nucleus. Because of the unique atomic vibratory patterns of electrons in different elements, only certain orbits are possible for a particular element.

Electrons always seek ground state, or the lowest possible orbit for an electron around its nucleus. To return to ground state, an electron must give off enough energy chunks to close the lower, ground-state orbit without disturbing the oscillating frequency. When this occurs light and other forms of radiation, such as heat, release, or become emitted by the electron, a kind of reversal in the absorption process. Therefore, energy may move in and out of an atom only in discrete packets that are uniquely suited to the frequency and orbit of each electron associated with that atom. Among other things, this is how we create colors. Certain substances, or atomic elements, interact with light in certain ways. Light links, so to speak, with the discrete molecular structures of materials like paint, dye, etc.

140

In the course of proving his theory, Planck concluded that the packets of discrete energy, which he called *quanta*, were limited in their discrete amounts according to the oscillating elements comprising the black body. Therefore, the energy contained in a quantum could be determined, because it is proportional to the frequency of oscillation of the blackbody radiation. He calculated the relative energy by multiplying the frequency of oscillation by a constant of 6.626E-34, known as Planck's constant. Planck's law of radiation therefore predicts the amount of energy at any given wavelength.

Planck's theory of blackbody radiation first appeared in 1900. Yet despite the logic and elegance of his formula (and because it contradicted popular classical wave theory of the time), quantum theory was not accepted until 1913, when the Danish physicist Neils Bohr demonstrated the practical application of Planck's theory to the spectral curves of stars. By measuring the relative energy output of light frequencies emitted by a star, and applying Planck's formula, Bohr was able to obtain the approximate temperature of a star, which, in turn, divulged a great deal of information about the star's age, luminosity, and ultimately its distance. Ten years later, the American physicist, A. H. Compton, proved the existence of quanta in electromagnetic radiation. However, it would not be until 1926 that quanta would obtain the name *photons*.

141

A significant characteristic of black body radiation is its emission curve, the relative energy emitted by a mass at different energy frequencies. Figure **14.6** is a representation of the Planck blackbody curve. A remarkable thing about the Planck curve is that it is always the same general shape for a black body regardless of the body's temperature. The peak of maximum intensity changes as the mass of the body changes, but the shape of the curve always remains essentially unchanged. The radiation peak of the Planck

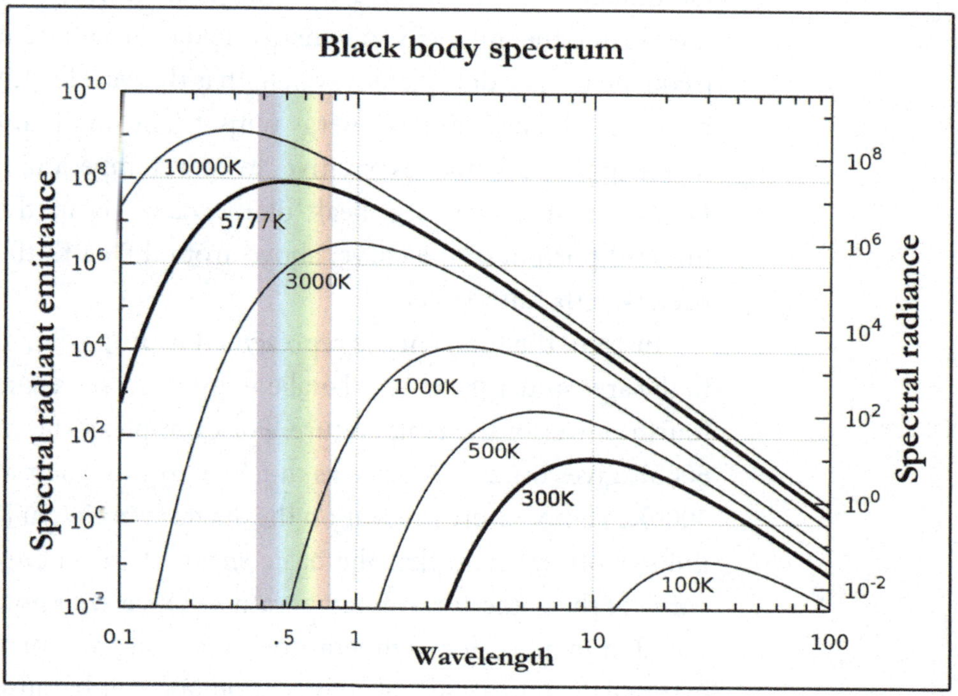

14.6. Planck's blackbody spectrum curve.

curve defines the dominant color of the radiation.

Knowing the dominant color of a star enables the prediction of its temperature because of the regular shape of the Planck curve.

The temperature of a star, in turn, is useful to calculate the star's mass. Beyond a temperature of about 900K, some portion of a star's blackbody radiation will always be inside the visible range of the Planck curve, even though the peak of the curve, in some cases, may be in the invisible parts of the spectrum, lying in either the ultraviolet or infrared portions. The peaks of the curves shift to higher frequencies as energy output increases. All stars produce energy across a broad spectrum. Stars like our Sun, for example, with its peak radiation at approximately 6000K, produce a large amount of energy in the visible light range.

The Planck curve traces the light emitted at different wavelengths. For a star of a given mass and temperature, the curve centers on one dominant wavelength.

Dominant color temperature is easily determined by measuring the intensity of energy taken at different frequencies. The dominant wavelength is the wavelength having the highest level of energy output. The dominant wavelength peak moves relative to star temperature. Below about 3000K, the peak of the curve is in the infrared portion of the curve; above around 8000K, the peak is in the ultraviolet.

In **14.6,** the total energy represented is proportional to the area under the curve. Total energy increases as the fourth power of the temperature. For example, a star at 6000K gives off $2^4 = 16$ times as much energy as a star at 3000K. Temperatures appear on the diagram in Kelvin K units. Notice, again, that the basic shape of the energy curve is the same for any temperature. Once we know the dominant color temperature of a star, we may extrapolate a great deal of information about it because the energy output of a star directly relates to its temperature

The colors of stars are telling us something about their temperatures. High temperature blue stars are substantially more energetic, thus potentially more luminous than low temperature red stars. Planck's law gives us relative energy output, or luminance. From this, much knowledge emerges about a star's life and ultimately its distance in the universe.

If dominant wavelength radiation has its peak outside the visible range, the star will still be visible but not in its dominant color. Except for bodies of the highest temperatures, the dominant wavelength peak for most stars is in, or close to, the visible region. However, visible observation of stars alone may not be sufficient to establish the Planck curve. We need spectrographic analysis to establish the full range of radiation, including the invisible ultraviolet and infrared energy.

142

Spectroscopy

Spectroscopy is the science of measuring the relative energy in a beam of light. Spectroscopic analysis of light from a star is a method to determine the intensity of the radiation and the characteristics of the elements producing it. Spectroscopic data is all that one needs to predict a great deal about a star.

Astrophysicists employ the science of spectrographic technology to identify the composition of matter in the universe by analyzing light and other types of electromagnetic energy. Until its discovery in the early part of the 19th century and the invention of the spectroscope, the only other technique available for studying matter beyond Earth was to gather up remnants of meteorites for chemical analysis. Figure **14.7** illustrates how spectroscopy works.

A spectrograph is a device used to photograph or map color spectra. It consists of a slit, a collimator or optical device to produce parallel light rays, a prism for dispersing light, a lens to focus the spectral image, and

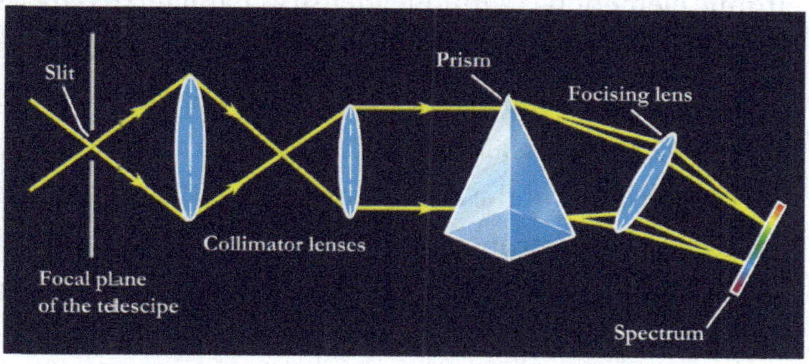

14-7. Spectrograph optics.

film to record the image. The photographic record produced by the spectrograph is a spectrogram. Pictured in **14.8** is a spectrogram of the element hydrogen.

The basis of Spectrography rests on the discovery by Isaac Newton who separated light by a prism into bands of color. A century after Newton's observation, a fellow Englishman, Thomas Young, showed that the different colors in Newton's spectrum were identifiable by

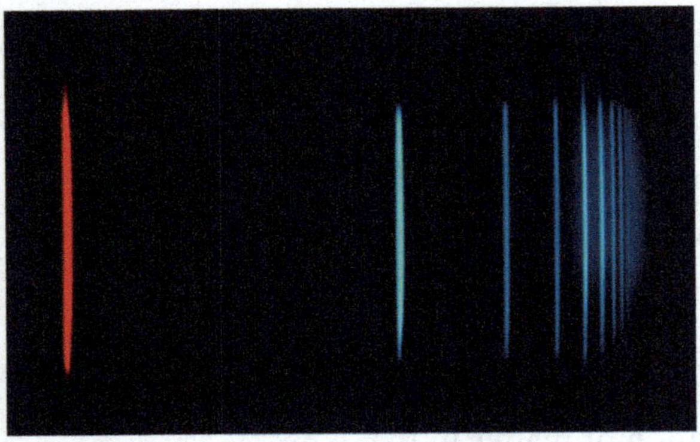

14.8. Spectrogram of the element hydrogen. Schawlow and Hansch, *Britannica*.

different wavelengths, setting the stage for connecting energy to color.

Soon after Young's discovery, the German physicist Joseph von Fraunhofer in 1814 constructed the first experiments to associate the lines of spectrograms with specific elements. He discovered, for instance, hundreds of thin dark lines interrupting the continuous spectrum of the Sun. Newton had not seen these lines in his experiment because he had used a circular pinhole to introduce light, where Fraunhofer employed a narrow slit. Still known today as Fraunhofer Lines, we now know that the pattern of lines observed in Sunlight identifies the various elements in the Sun's atmosphere.

By 1868, physicists Gustav Robert Kirchhoff, R. W. Bunsen, and the Swedish scientist Anders Jonas Ångström conclusively demonstrated that each type of atomic element has a unique identity characterized by its own line spectrum; like a thumbprint, no two elements give off the same spectrum. Armed with blueprints of the known elements, scientists quickly were able to identify the composition of the Sun's atmosphere, thus began the new science of astrophysics. By 1925, spectral chemical analysis was a common practice extended to stars beyond the Sun.

Today, the science of spectroscopy in its many forms and applications is the primary method by which planetary atmospheres, galaxies, stars, nebulas, and other celestial objects undergo analysis for their physical and chemical properties. The application of modern spectroscopy also predicts the radial motions and distances of bodies in the universe.

Meanwhile, the results of spectroscopic analysis determine where a star belongs on the main sequence H-R curve. A star's energy wave spectrum reveals the composition of the star and its dominant wavelength, or color, which in turn gives the temperature of the star from the Planck blackbody curve. Once we know the dominant wavelength of a star, this single number is enough to establish the entire Planck curve and to identify the star's place as an H-R main sequence star. Consequently, spectroscopic data alone predicts the mass and density of a star, thus establishing its overall energy level. Since the energy of a star manifests in its luminance, its magnitude calculation is possible, and the inverse square law applied to magnitude predicts the star's distance.

143

Stars of variable brightness

Variable stars are stars that at first shine brightly, and then become dim before growing bright again. Stars mysteriously change in their appearances, something known for thousands of years. There is evidence that the Babylonians knew the star Mira, for instance, to be a variable star five thousand years before its official discovery in the 16th century. The Chinese and Koreans knew about Mira, as did the Greek Hipparchus, most likely.

Mira disappears periodically, only to reappear predictably later on. Mira changes from a visibility of second magnitude to below sixth magnitude when it disappears from view completely, only to reappear again, increasing in brightness to peak again at second magnitude. Early astronomers could only marvel at this phenomenon, having no idea what might cause it. Mira is Arabic for *miraculous,* or *wonderful,* a name given to it in the 17th century, probably in appreciation for its then unique apparition.

Mira has a primary cycle repeated every 331 days, with a secondary cycle occurring every six months, and another overall change about every eleven years. During its minor six-month cycle, Mira changes in brightness by a factor of 100 times, and during its major 331-day cycle, it grows to an absolute brightness that is 250 times as bright as the Sun.

144

Mira, known also as the star Omicron Ceti, lies approximately 400 light-year s from Earth. Located in the constellation of Cetus the Whale, Figure **14.9** is a picture of this star. By comparison, if superimposed on the solar system, Mira would reach well beyond the orbit of Mars. It is on average more than 300 times the diameter of the Sun—on average because it changes size.

14.9. Mira. Visible as a dot in the bulb shape at right, Mira is speeding through space at 130 kilometers per second, faster than the speed of a bullet. Brighter stars in the foreground of the photograph are closer than Mira. NASA Galaxy Evolution Explorer.

Mira was resolved in 1995 by the European Space Agency's Faint Object Camera to show a companion star, Mira B. Mira B is a hot, white dwarf star believed to play a role in Mira's irregular harmonic pulsations. The Hubble Space Telescope confirmed Mira B's discovery in 1997. For the first time, too, Mira proved to be not round but of irregular shape, like a football. The Hubble revealed that matter from Mira A flows outward, drawn toward Mira B across an enormous distance. The two stars are more than 70 times as far apart as the distance from Earth to the Sun.

Mira owes part of its change in luminosity to changes in its size. During its primary 331-day cycle, it more than triples in size, growing from an initial diameter roughly 200 times that of the Sun to a maximum size 700 times the diameter of the Sun. Its oscillations are due also, in part, to accompanying temperature changes. Mira alternates in temperature from a visible reddish-orange color at 2500K, down to a redder, less visible 1900K. These changes when taken together, change of size and fluctuations in temperature, account for Mira's periodic change in brightness. Mira expands from a relatively small, cool star to a much larger, hotter one, all in predictable periodic cycles.

Cycles of change in stellar bodies are common. Though not nearly as unstable as a variable star, our own Sun goes through various cycles of sunspots. It has an especially active eleven-year cycle that last peaked in 2014. Giant flares of energy extend deep into space and may cause, among other things, dazzling northern lights displays as far south as New Mexico. The Sun, however, despite its impressive activity, has not yet entered the dramatic oscillating phase seen in Mira and other kinds of variable stars.

145

Cepheid stars

Over the decades, the phenomenon of periodic star oscillation gained better understanding. Differences surfaced in oscillating star behaviors. The differences appeared to follow definite patterns. Cepheids, for instance — another class of oscillating stars different from Mira-type stars — completed their cycles in a few days. Mira-type stars, on the other hand, varied over a period of many months or even years. Slower oscillating, Mira-type variables identify as less massive stars. Cepheid stars, on the other hand, are usually associated with giant stars and supergiants. Cepheids oscillate much more rapidly than Mira variables, from one to two days on the fast side, to around 100 days at the slower end. Their brightness diminishes gradually before rapidly building back to maximum luminosity; that is, they accelerate in brightness rapidly, and then slowly return to normal before repeating the cycle. Delta Cephei, for instance quickly reaches a peak luminosity 3,000 times that of the Sun, then gradually tapers off over a period of 5.4 days. It changes from a maximum magnitude of +3.6, down to a minimum brightness of +4.3 magnitude.

In the year 1784, the young English astronomer John Goodricke, pictured in **14.10,** first recorded Delta Cephei. We know of thousands of Cepheid stars all catalogued since Goodricke's discovery, along with many Mira-like stars. Astronomers are now able to account for a host of other star types that exhibit characteristics similar to, but unique from these two original variable types. The North Star Polaris turns out to be a Cepheid star with an oscillation period of just less than four days. However, its brightness changes so slightly that its variations are difficult to see with the unaided eye.

14-10. Portrait of John Goodricke. Painting by James Scouler, c. 1785. Royal Astronomical Society of London.

The classification of Cepheid-like stars has expanded significantly since John Goodricke first recorded the variability of Delta Cephei. Cepheid stars like Delta Cephei that are predictable in their period-luminosity relationships we call Classical Cepheids. The cycles of change in these stars may vary from star to star from 1.5 days to more than 50 days but each individual star is consistent in its period of oscillation. Classical Cepheids are usually younger stars known generally as Population I Cepheids. Population II Cepheids, on the other hand, are older and have much shorter-period variables with oscillations of less than one day. These faster oscillating stars we sometimes call cluster-type variables, or RR Lyrae-type variables. Their luminosity-to-oscillation relationship is somewhat less dependable than classical types like Delta Cephei. They are, nevertheless, good distance estimators, since all such stars tend to be the same brightness wherever they are located in the universe. Therefore, if they have the same intrinsic brightness, then observed differences in magnitude must be due to distance.

146

In 1912 the American astronomer Henrietta Swan Leavitt, summarizing work she began in 1908, revealed that the oscillation period of a Cepheid star directly corresponds with its intrinsic luminosity. Cepheids with longer periods of brightness oscillation change are proportionally brighter than less luminous ones that pulsate faster. Put another way, the more luminous a Cepheid star is, the longer is its period of change, and vice versa. Her landmark work with Cepheid stars was a turning point in cosmic exploration. She appears pictured in **14.11**.

14.11. Portrait of Henrietta Swan Leavitt (1868-1921). Unknown photographer. American Association of Variable Star Observers.

Once Leavitt worked out the formula of the period-luminosity (P-L) relationship of Cepheids, a new tool for estimating distances to cosmic bodies was on the horizon because from the period of oscillation of a Cepheid variable, mass, luminosity, and absolute magnitude may be calculated. Leavitt catalogued hundreds of Cepheid variable stars in the nearby Magellanic Clouds and found that the longer the period of oscillation, the brighter the star, thus paving the way for new discoveries about the size of the universe.

Leavitt grew up in Massachusetts, went to school at Oberlin College, graduated from Radcliffe, and spent several years as a volunteer at the Harvard Observatory before her appointment to the observatory's permanent staff, eventually becoming Head of the Department of Photographic Photometry, and spending the rest of her life at Harvard.

Like John Goodricke, discoverer of Delta Cephei, Leavitt was deaf. Illness resulted in almost complete loss of hearing shortly after college. Yet, she continued to exhibit a single-minded perseverance of study and self-application to her work. Her work on star magnitude calibration was for years referred to among astronomers as the Harvard Standard.

Leavitt never had the privilege of following up her work. As one of several people — mostly women — hired as computers, the director of her observatory gave her no analytical assignments. The director believed that her job and that of her female colleagues was to collect data, not to analyze and interpret it. This task he assigned to Harlow Shapley, remembered, among other things, for his expansion of the known universe by identifying the center of the Milky Way Galaxy. Shapley used Leavitt's technique to help make that discovery. Meanwhile, Edwin Hubble also relied on Leavitt's work to show that the Andromeda Galaxy was outside the Milky Way, thus marking another major expansion discovery. Therefore, two of the great leaps into the universe trace, at least in part, to Henrietta Leavitt. Yet despite the enormity of her contribution to astronomy, Leavitt never received the acclaim she deserved. At the time of her death, she had cataloged some 2,400 variable stars, half of all the variable stars known at that time.

She died in 1921 of cancer at the age of 53. When she died, she was working on a new photographic stellar magnitude scale. Colleagues tried to nominate her for the Nobel Prize posthumously in 1926. However, the prize committee does not award the Nobel Prize posthumously. Therefore, the only lasting tribute to her is a lunar map feature that bears the Leavitt name to honor her and all the other deaf men and women who have worked in astronomy.

Leavitt's technique was a simple one. The target of her observations was the Magellanic Clouds. The Magellanic Clouds are those irregular galaxies belonging to the Local Group, meaning they are neighbors to the Milky Way Galaxy. Visible only from the southern hemisphere, the clouds first had their discovery by the ship crew of Ferdinand Magellan, in Magellan's great circumnavigation of the world in the early 16th century. Observers referred to them as clouds because of their hazy, nebulous appearance. Probably formed after the Milky Way, these galactic companions comprise both new stars and star clusters, as well as substantially older stars. The Clouds are 180,000 to 200,000 light-years away.

Not knowing even in her time that the Magellanic Clouds were actually galaxies, Leavitt nevertheless reasoned that stars in each Magellanic Cloud would be at approximately the same distance. Thus, the apparent magnitudes of stars in one of the clouds would be precisely in the same relationship as their absolute magnitudes. She then was able to apply the period of oscillation of each Cepheid variable to its unique magnitude. Still, Leavitt could only measure apparent magnitude, because the distance to the Magellanic

Clouds stood yet unknown. Full discovery required a benchmark reference Cepheid star needed to apply Leavitt's findings to the prediction of stellar distance. That came a year later in 1913 when Danish engineer Ejnar Hertzsprung used Cepheid variables to determine the distance to the Small Magellanic Cloud. This was the first distance determination of an object outside our galaxy.

Hertzsprung had shown in 1905 that the colors of stars relate to their brightness. It was Hertzsprung's work with Cepheid variables eight years later, coupled with Leavitt's Cepheid luminosity-period law that established the Cepheid luminosity scale, and opened the way for measurement of intergalactic distances. It is irrelevant to his accomplishments but interesting to note that Hertzsprung was a chemical engineer who, like other great astronomers before him, had no formal education in astronomy.

The magnitude scale of Leavitt's Period-Luminance diagram, once calibrated to the Hertzsprung data, resulted in a revolutionary new tool for determining distance. Added to the stellar parallax and magnitude techniques for stellar distance prediction, estimates are that Leavitt's discovery significantly enlarged the potential size of the known universe. Thought to be about 100 light-years at that time, this science enlarged the universe by a factor of at least 100,000, creating at the same time a new yardstick of the universe and introducing the earliest method for the determination of extra-galactic distances.

From the rate of change of a Cepheid variable, it is possible to estimate its luminosity, and knowing the luminosity of a star makes it possible to calculate its distance by its apparent magnitude. Application of the inverse square law to the magnitude of a Cepheid star enables prediction of its distance to a high degree of accuracy. Edwin Hubble used this method in 1924 when he discovered Cepheid variable stars in the Andromeda Galaxy to show that Andromeda lay far beyond the Milky Way Galaxy. It lay much further out than previously thought, and at a distance that revolutionized the prevailing perception of the size of the universe.

Both Cepheids variable stars and Mira variables have played a key role in calculating the size and composition of the known universe. Harlow Shapley based his work in discovering the center of the Milky Way Galaxy on Cepheids, and, of course, Hubble's discovery of Cepheid variables in Andromeda began a new wave of space exploration that forever changed the perception of the scale of the universe.

David Vinckboons. *Petrus Plancius Instructing Students in the Science of Navigation*. bef 1629. Black and brown ink and wash, 10 ¼ x 11 ¼ in. (26 x 28.6 cm.) Frits and Rita Markus Fund, 2003. Metropolitan Museum of Art.

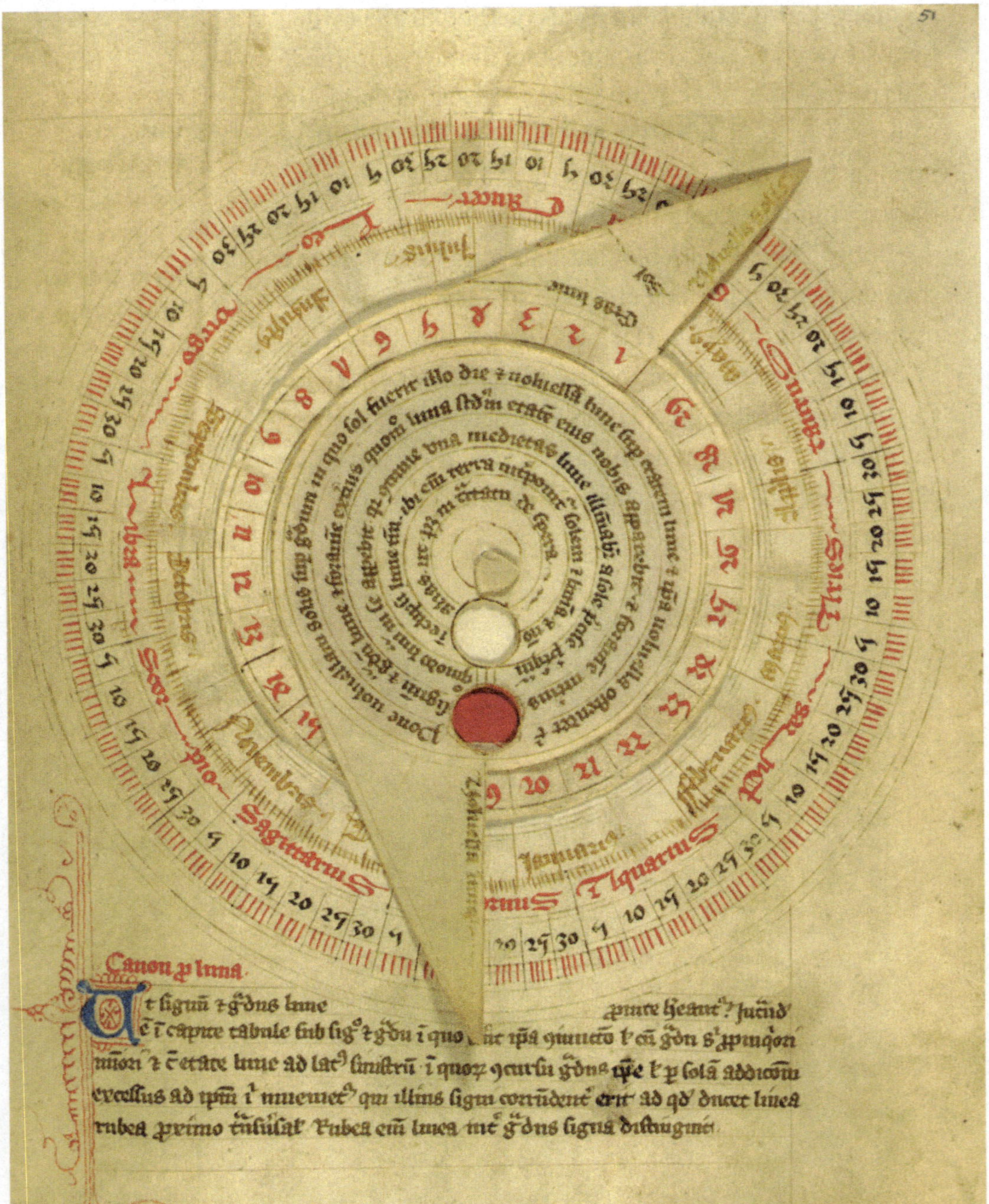

Astronomical Table with Volvelle. ca. 1386.
Pen and black ink and tempera, 8 ⅜ × 6 in. (21.3 × 15.2 cm.)
The Getty Art Museum.
Based on the Toledan Constellation Tables, ca. 1080. A moveable volvella predicted the position of the sun and moon in the zodiac. Copernicus used the tables in his model of the heliocentric universe.

MANITOU OF THE DEEP FIELD

The cosmos is all that is or ever was or ever will be.

Carl Sagan.
Cosmos, 1980.

Hendrick Goltzius. *Jupiter Taking Counsel from the Gods about the Destruction of the Universe.* 1589.
Engraving on laid paper, 6 5/16 x 13 7/16 in. (16.1 x 34.1 cm.)
National Gallery of Art.

XVI

DEATH OF A STAR

Stars fall into three broad size categories: small, mid-size, and large. In each of these categories, stars behave differently during their lifetimes, progressing through various stages of change depending on their initial masses. The greater portion of the universe is illuminated by mid-size, main sequence stars, most of them very much like our own Sun, some hotter some cooler. The 10 percent or so of stars not on the main sequence are stars in transition.

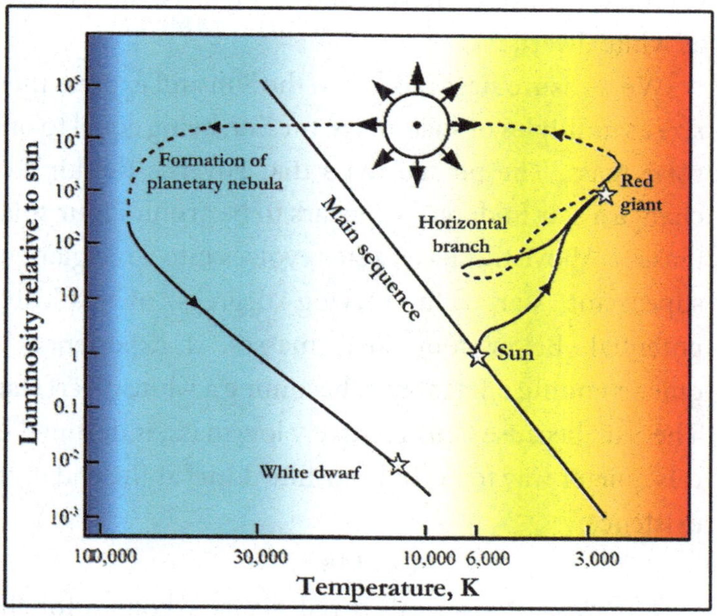

15.1. Evolution of the Sun at the end of its existence.

At some point in a star's life, nuclear fusion ceases. As a star consumes its nuclear fuel, changes in it begin to occur. Eventually a star will leave the main sequence to commence the final stages of its existence (**15.1**). It becomes a white dwarf, red giant, or red supergiant star.

Which of these three types of stars it becomes and what happens next depends on its size.

This diagram in **15-1** illustrates the typical evolution of a star the size of the Sun. Larger or smaller stars may evolve differently. When the fuel of a star finally exhausts itself at around 10 billion years for an average star like the Sun, it leaves the main sequence to begin the last phases of its existence. A star the size of the Sun will at some point, as it depletes its nuclear fuel, leave its main sequence appearance to become a red giant star. As the star sheds its outer layers, it enters a phase called planetary nebula. Hotter, more massive stars explode as supernova, but a star the size of the Sun usually does not explode at the end of its giant phase as larger stars do. Sun-like stars, once free of their outer layers of gasses, eventually cool down to become faint, dense stars known as white dwarfs.

We measure stars relative to the Sun and express their size as multiples of solar mass, the Sun being equal to one solar mass. The pivotal mass that governs the kind of change a star undergoes appears to be around four solar masses. Above this level, a star evolves into a red giant or supergiant star with varying degrees of explosive potential. Below four solar masses, it experiences a quieter ending on its way to becoming a white dwarf star. The Sun, because of its relatively low mass, is destined in this quieter way to become a white dwarf at the end of its existence.

148

Red giant stars

The class of stars called red giants accounts for the transformation of mid-size stars between four and eight solar masses, that is, four to eight times the size of the Sun. A star this size will collapse at the end of its life, sending out a brilliant burst of energy, leaving a white dwarf star ultimately to cool and become a densely packed black, invisible body with a strong gravitational field that belies its shrunken size. Since hotter main sequence stars are the more massive stars, predictably they are the ones to watch for red-giant activity.

Surprisingly, larger stars have much shorter lives than cooler stars, because they consume their nuclear fuel at a much faster rate. Consequently, there are not as many of them. A larger star may last no more than 10 million years at most, as compared with smaller stars that may burn for 10 billion years or more. Sirius is one example of a fast burning star. It is burning up at a rate 20 times that of the Sun. Nearly twice as large in diameter as the Sun, and 23 times more luminous, with a mass 2.4 times as great, Sirius will live barely one-tenth as long as the Sun.

As a star exhausts its fuel, its nuclear activity becomes insufficient to sustain its physical integrity; it becomes unstable. Its distribution of mass begins to change, and as it does, so does the composition of its atomic elements. A star that has used up its nuclear fuel goes through a spectacular phase of ejecting its outer layers. The inner core of the star separates from its outer atmosphere. As the separated outer layer expands, it begins to cool. At the same time, gravity releases its hold on this expanding, less massive outer shell. The exposed inner core, by contrast, contracts around its center of gravity and becomes denser, widening in the process the separation from the outer layer, and further exposing the extreme temperature of the core. Measurement of the temperature of a disintegrating main sequence star is of its outside shell surface temperature; the inside temperatures of stars are unknown. Nevertheless, some astronomers have speculated that a large main sequence star may have internal temperatures in the millions of degrees. Such a theory is reasonable. In our experience, Earth's average surface temperature of around 70 degrees Fahrenheit belies an inner core temperature 100 times that amount.

Dying stars become enormous before they finally disintegrate. As gravity releases the less dense, outer layer of an expanding star, the star's expansion causes it to occupy an ever-enlarging space as it spreads out. A red giant star expands to several hundred million kilometers in diameter from its normal three million kilometers, or so, as a main sequence star. Consequently, a red giant star will appear very brilliant. The Sun, a million times as dense as the outer layer of a typical red giant, is only one millionth as bright as one of these giant stars.

149

Arcturus and Capella

Two stars that are now entering the final phases of their stellar lives are Capella, located near the celestial north pole in the constellation Aurigae (the charioteer), and Arcturus located in the constellation Boötes (the herdsman). These two yellow giant stars, respectively the sixth and fourth brightest stars in the sky, have already begun the process of fusing their inner cores, prelude to casting off their outer helium shells.

Capella, which is 46 light-years away, is spectrally more like the Sun. However, unlike the Sun, it is already expanding, as dying stars do. Capella, not one but two yellow giant stars orbiting very close to each other and unfortunately not resolvable by telescope, is 16 times the size of the Sun and 150 times brighter. Capella is the star nearest to the North Pole Star and thus is an important navigational reference because it is always visible from the northern latitudes.

Arcturus, on the other hand, is 20-25 times the Sun's radius and 115 times as bright. Closer than Capella at 37 light-years, Arcturus classifies as an orange giant star, meaning it is probably further along than Capella toward red giant status and its final days. Arcturus appears in the spring and early summer months. Look to the southwest beyond the arc of the Big Dipper; follow the "arc to Arcturus" the orange star.

Among peoples in the northern latitudes, Arcturus, like the star Sirius, is the focus of metaphysical and religious speculation, as it has been for centuries. From a northern vantage point, Arcturus is not the fourth but the second most visible star in the sky, behind Sirius; thus its heightened importance to primitive cultures of northern latitudes. Biblical writings mention Arcturus as the Star of Job. Homer mentions it, too, in his *Odyssey*, and it was a topic of Arabic and Chinese mythology, as well as North American Indian lore. Some of its references are at least 2,500 years old, giving it a distinction of being one of the oldest named stars.

The 20th century psychic Edgar Cayce also mentioned Arcturus as the place of an advanced civilization. Needless to say, the Cayce readings, because of the credibility given to Cayce by learned professionals in many fields, spawned entire generations of imaginative writers who believe that the Cayce vision of the Arcturian 'fifth dimension' sanctions the many popular literary and cinematic productions of aliens fitting the classic Arcturian description, not to mention a whole language of Arcturian characteristics.

150

Nebulae

What happens to ordinary stars below four solar masses like our Sun? The deaths of small stars are far less cataclysmic than giant stars, although the process of extinction is similar. The Sun will never become a red giant star because it lacks sufficient mass; it nevertheless will undergo a similar, though less dramatic, expansion process at the end of its life. Hence the prediction by astronomers that the Sun will expand at the end of its normal existence to engulf the inner planets is based on the predictable death sequence of an exhausted star.

Figure **15.2** illustrates a pair of relatively small stars, each one undergoing separation of its core and atmosphere. The Cat's Eye Nebula (top), three thousand light-years distant, is a star in the complex final phases of its existence as a star. Similarly, time is running out on the star of the hourglass-shaped nebula. These stars are destined to become white dwarf stars, much like the Sun will at the end of its existence.

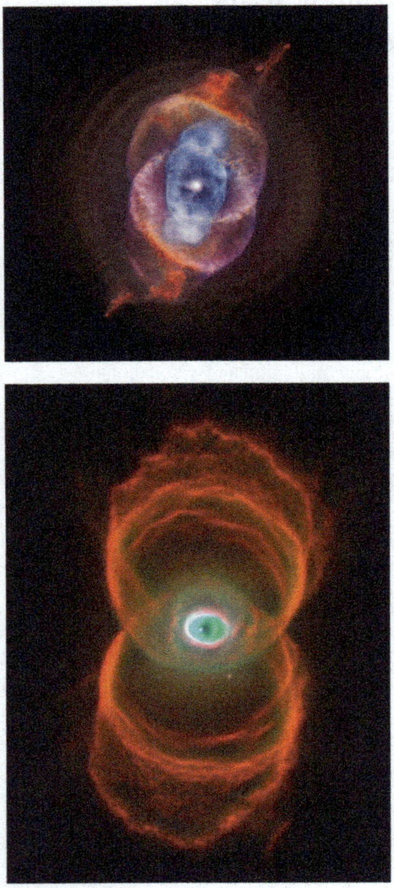

15.2. Cats Eye and Hourglass Nebulae. Planetary nebulae have at their center a core of the original star in the form of a white dwarf star. Different hues depict various gaseous elements as they separate from the central star. The nearest planetary nebula is about 400 light years away. NASA.

As the fuel of a smaller, main sequence star like the Sun is exhausted, the core and outer layer of the star separate, just as they do in larger stars. The star sheds its outer layer into a nova, or, in the case of the smallest stars like our Sun, a planetary nebula, so called because at one time astronomers believed it was the stuff from which planets formed.

The shedding of the outer layer of an ordinary small star may recur more than once, until eventually, the star's inner core is all that is left. There is no explosion, at least not on the scale of a supergiant star. The removal of the outer layer of a dying, small star is analogous to puffs rather than explosions.

A star can eject its outer layer in many ways, resulting in a huge variety of unusual forms and shapes. The most famous planetary nebula is the Ring Nebula, shown in **15.3**. Because planetary nebulae are the remnants of dying Sun-like stars, they are numerous. After burning for billions of years, these stars will continue to glow as remnants for several thousand more years before their gas rings dissipate or assimilate into other cosmic bodies, sometimes to be recycled as new stars.

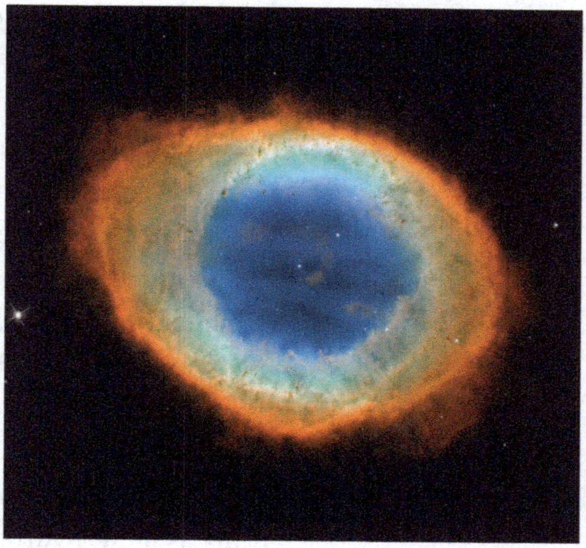

15.3. Ring Nebula. The Ring Nebula is the most famous of all planetary nebulae. Located 2,000 light years from earth in the direction of the constellation Lyra, it appears here approximately in its true colors. NASA, ESA, and the Hubble Telescope.

Nebulae forms appear as rings of gas. They are, in fact, spherical cocoons that only look circular because the relative density at the outer edges aligns with our point of view, like looking through the atmosphere of Earth at a low angle.

The French astronomer Charles Messier who in 1781 compiled the Messier catalog, a list of 103 star clusters and nebulae objects, first adopted the term nebula. The Messier catalog is the source of the *M* prefix that accompanies the identification of many cosmic bodies. As mentioned before, the Andromeda Galaxy, for instance, catalogs as object M31, the 31st object recorded by Messier. The Crab Nebula is M1, the first object cataloged. Nebula is Latin for *cloud*, which Messier believed aptly described a class of amorphous stellar shapes he called nebulous objects.

151

The death phase of a red giant star is a nova. Nova is Latin for "new star," a term coined by Tycho Brahe when he mistook the explosive death of a supergiant star for the birth of a new star. A nova star occurs somewhere in our vicinity about every 10 years. Several occurred in the 20th century alone, some especially spectacular. Nova Aquila, for example, known also as the Nova of 1918, shined clearly in the nighttime sky. Within hours it became as bright as the star Sirius, growing in brightness by a factor of 100,000 in just six days, only to gradually diminish in magnitude over the next seven years before fading out of sight. Mysteriously, Nova Aquila appeared the same year as the great flu epidemic that killed 20 million people worldwide, thus heightening superstition that cosmic events tie to events on Earth.

Coincidentally, one of the more recent nova recorded in 1999 also was located in the constellation of Aquila. Not quite as bright as the 1918 Aquila Nova, this one, called Nova Aquila No. 2 reached a visual magnitude of +3.0. The estimate of the 1918 nova was magnitude -1.1.

Astronomers believe that some novae originate from an interacting binary star system, one star of which is a white dwarf star. The white dwarf pulls material from its companion star. If this hydrogen-rich accretion has sufficient mass, it ignites, causing the outburst of luminosity seen as a nova. This phenomenon can repeat itself more than once, causing a star periodically to flare up as new hydrogen material is attracted. There are several examples of recurrent novae in the Milky Way. However, if all novae require a multiple star system, then for that reason alone time will spare the Sun from a too cataclysmic ending because it has no companion star.

152

White dwarf stars

The collapsed inner core of an ordinary star is a white dwarf. Since first discovered in the 19th century, hundreds of these stars are observable. The globular star cluster M4 is home to a large concentration of cooling white dwarf stars. Astronomers hope to learn more about the age of the universe by studying the decaying energy of these stars as they cool and fade away, because each one was once a main sequence star, possibly very much like our Sun. White dwarfs are highly concentrated masses, typically about the size of Earth (**15.4**). A teaspoonful of the matter of a white dwarf of around one and one-half solar masses would weigh five tons, or like compressing an elephant into a space the size of a sugar cube.

In the year 1896, astronomers discovered the star Procyon in the constellation Canis Minor to have a companion star, a star calculated to be more massive than the Sun, yet no larger than Earth. That star,

15.4. Scale of the white dwarf star Sirius B compared to earth. A typical white dwarf star is about the size of earth. Its evolution is like compressing the Sun into a volume the size of the earth. Consequently, a white dwarf star has an extremely concentrated mass with a very high temperature, estimated above 200,000 degrees. ESA and Ohio State University.

designated Procyon B confirmed the existence of white dwarf stars after more than 50 years of mostly theoretical claims. A preliminary discovery made some 35 years earlier of a companion star near the star Sirius went unnoticed, not understood then to be a white dwarf. That star, known as Sirius B, we consider the first white dwarf star detected, although no one knew it at the time. Sirius B is a member of the Sirius binary star system. Together, these two discoveries: Procyon B and Sirius B, paved the way for the exploration of a whole new class of stars and added the final phase in the evolution and death of a star. Because of its relatively small size and visual obscurity next to Sirius, together with its large mass and strong gravitational field, Sirius B is also the first discovery of so-called dark matter, because detection of its energy field occurred long before its presence had telescopic confirmation.

153

Sirius B

Sirius B is possibly one of the youngest white dwarfs in the sky. This belief comes from ancient literature in which Sirius appears to have changed from red to white within a span of 800 years. Ancient classical writers refer to Sirius as red. Ptolemy himself classified it as fiery red. However, by 925 A.D. Sirius the white star had replaced references to Sirius the red star in the literature. This change of description makes sense if Sirius B was a main sequence star in its final stages of life. Sirius B would have been unknown at that time. When it expanded into a glowing red giant, many times as large in diameter and brightness as at any point in its existence, the redness of Sirius B would overcome the brightness of the main star

15.5. Typical size comparison between the white dwarf Sirius B (center), its companion Sirius A (left) and the Sun (right). Photo by R. J. Hall.

Sirius, thus causing early astronomers correctly to call Sirius a red star. (**15.5**)

On its way to becoming a white dwarf star, Sirius B would first have expanded into a brilliant red star, bright enough to be visible perhaps even during the day. No one knows when this may have occurred. There are, however, many references in the ancient literature to Sirius the red star. Astronomers today dismiss these sometimes-conflicting descriptions of the red Sirius as errors of observation. It is conceivable nevertheless, that Sirius and Sirius B together once appeared as one very bright star. The explosion of Sirius B would easily have altered the perceived color of Sirius before Sirius B contracted into its less visible white dwarf phase, leaving the brighter blue-white Sirius as a somewhat diminished but still a brilliant lone harbinger of dog days. In such a scenario, Sirius would have appeared white, then red, then white again, over a long period.

The implausible part of such a scenario is that it seems unlikely that the red expansion phase of Sirius B, which would normally be measured in thousands if not millions of years, would die out in a mere 800 years, unless it was observed in its final nova phase. In which case, the nova would likely have happened in months or years and not the decades or centuries that the ancient literature would seem to suggest. It is possible that Ptolemy and those before him saw the star at the end of its red phase and witnessed the contraction of a white dwarf star. This could account also for the sun-like qualities assigned to Sirius by the ancient Babylonian and Greek astronomers. As the redness of Sirius B faded, the whiteness of the primary star Sirius would have again asserted itself. Most astronomers do not think so, pointing instead to the more likely explanation that the classical literature is wrong, citing conflicting writings that tend to contradict the fiery red description. Ptolemy, they say, probably chose as his descriptive point that moment when Sirius rose above the horizon, thus causing the star to appear red just as the Sun appears red in the light of dawn or dusk. Just as we see Sirius as white, even though it is actually blue, for the same reasons of atmospheric interference, Sirius the white star could have appeared at certain times of the day in the morning and evening as Sirius the red star.

154

The Dogon legend

Sirius B appears as one star with Sirius A to the naked eye. Yet, Sirius B has been the subject of mysterious tales going back perhaps 5,000 years. The Egyptians, for example, allegedly knew about the gravitational dynamics of the orbiting Sirius B in 3200 B.C.E., long before its official discovery in 1862.

Meanwhile, the story of the African Dogon tribe of Timbuktu, Mali, is even more fantastic. They claim to have knowledge of the specific qualities and behavior of

Sirius B, including its orbiting period and the fact that it is a very dense, heavy star. Their knowledge, according to tribal claims, passed from generation to generation over many centuries; yet, science has only recently confirmed the nature of Sirius B. According to the Dogon story, alien space travelers came from a planet in the Sirius system, a planet yet undiscovered, to appear on Earth to the human race with angel-like purposes. Many details of the alleged alien visit also echo the legends of the Christian epics. Today anthropologists attribute the Dogon myth to concoctions of 19th and 20th century contact with terrestrial missionaries rather than alien travelers. Others believe it is an outright hoax. Still, there are details not fully explained, and cults have sprung up around the world to promote the Dogon legend. Perhaps someday we may discover that there really is a Sirius B planet that now only the Dogon people seem to know about. If it exists, it will be a world very different from our own; white dwarf stars spin at high speeds. Sirius B, for example, spins at an impressive 23 times a minute. The Sun, by comparison, turns on its axis once every 27 days on average. On average, because the Sun is a gaseous object so its equator actually rotates faster — 24 days — than its pole regions at 30 days.

155

New white dwarf stars are very hot, many times hotter than the Sun. Figure **15.6** shows a white dwarf star with one of the hottest temperatures ever measured, confirmed to be 200,000 degrees centigrade, more than 30 times hotter than the Sun's temperature. The intense heat of this white dwarf makes it 250 times brighter than the Sun. High temperatures result when the hydrogen of a once main sequence star converts to helium as it goes through its dying stages. Helium, in turn, converts to carbon, the byproduct of which is heat. Instead of radiating energy because of its nuclear power, the star now glows from its residual heat. Finally, it cools,

15.6. NGC 2440 in the Milky Way 4,000 light years from earth. The white dwarf star at the center of this planetary nebula has one of the hottest surface temperatures yet measured at 400,000 degrees Fahrenheit (200,000 degrees Celsius). Photo by K. Noll, NASA, ESA.

becoming dimmer in the process until it disappears into the darkness of space. The universe abounds in small, highly concentrated, massive bodies, many of which have gone dark and are no longer a part of the illuminated universe.

156

Black dwarfs

A black dwarf star is a star that has gone out. When the inner core of a white dwarf cools, it gradually loses its power of self-illumination — it goes out. At this point, the star is no longer a white dwarf but a black dwarf, retaining its density but having no visible characteristics. It disappears into the blackness of space.

We know a great deal about red giant stars and white dwarfs. However, we know virtually nothing about black dwarfs because they account for zero illumination detected only through indirect means. They are too small to be detected by telescope and do not radiate energy. Their only means of detection is by gravity and the effects of their strong gravitational force on other bodies. Black dwarfs are about the size of Earth, perhaps up to 32,000 kilometers in diameter but with a condensed mass whose field yields gravity 100,000 times that of Earth. Some of the many orbiting multiple star systems likely include one or more black dwarfs.

Black dwarfs theoretically are scattered throughout the universe as remnants of once main sequence stars. Our Sun likely will be just such a remnant itself five billion or so years into the future, a burned out ember like a giant chunk of spent coal.

The Elliptical Galaxy Messier 87. At the galaxy's center is a supermassive black hole 55 million light-years from Earth, and dramatically different from black dwarf stars. Spitzer Space Telescope. NASA/JPL-Caltech/IPAC/ Event Horizon Telescope Collaboration.

XVII

SUPERNOVAE AND NEUTRON STARS

A dying star of up to eight solar masses in size falls in the red giant class of stars. Dying stars above eight solar masses are red *supergiants*. At the end of its life, a red supergiant star has sufficient mass in its separated outer layer to become one of the great stellar events of the universe, the great supernova. A supernova is a cataclysmic explosion that results in the creation of a tiny, super dense star of nearly indescribable gravitational force.

16.1. Alpha Orionis, called Betelgeuse, is the brightest star in the winter constellation Orion. Xavier Haubois Paris Observatory.

There are a relatively few supergiant stars known to be in existence because the life of a red supergiant is comparatively brief. Although there are few such stars in the sky, the star Betelgeuse is one such body. It appears in **16.1**. At 650 times the diameter of the Sun, Betelgeuse appears to be a classic example of a red supergiant star. Compared with our solar system, if it replaced the Sun, the atmosphere of Betelgeuse would extend well beyond the orbit of Jupiter. Betelgeuse is about 600 light-years away

Betelgeuse contains a mass equal to 20 solar masses, burning up at a rate 500 times as fast as the

Sun. It varies in brightness from 700 to 14,000 times more luminous than the Sun, changing erratically over a period of about 5.7 years, perhaps already signaling by its expansion and contraction the approaching climax of its existence.

Halfway around the Zodiac from Orion and the star Betelgeuse is the constellation Scorpius, home of Antares, another red supergiant star. As Orion and Betelgeuse set in the winter sky, Scorpios and Antares rise in the summer sky. Antares is named for its rival appearance to the red planet Mars. Antares literally means *anti Aries.* Aries was the Roman equivalent of the Greek god Mars, god of war.

Like Betelgeuse, Antares is a huge star 700 times the diameter of the Sun. Already, Antares has begun its final stages of life by expelling gas from its surface, aided in this process by a blue companion star, Antares B, which is drawing off the gases from Antares A. Both Betelgeuse and Antares will explode as supernovae sometime in the not too distant future, probably sometime within the next 10,000 years.

158

The supernova of the year 1054

The explosive energy of a supernova star makes it so bright that it is visible during daylight. History has recorded seven visible supernovae seen from Earth in the last two thousand years, five in the last one thousand years. The earliest supernova recorded appears to have occurred in 185, followed by another in the year 393 and still another in 1006.

The most famous supernova, however, occurred on July 4, 1054, when the Chinese recorded one sufficiently bright to be called a *guest star*, so called because it could be seen plainly during the day at dawn and dusk sitting alongside the planet Venus. It remained visible in the sky for two years before fading out of sight.

Native American Indians, according to archeologists, recorded the same supernova. The petroglyph in **16.2** is allegedly an Anasazi Indian record of the same

16.2. Chaco Canyon petroglyph. The Anasazi residents of Chaco Canyon drew a celestial event of some kind in proximity to the moon. The diagram on the floor below the petroglyph may represent Halley's Comet, which appeared just a few years after the 1054 supernova. The tail of the comet trails away to the right. Photo by D. Heyser

supernova event witnessed by the Chinese in 1054. The Anasazi who made the drawing could easily have recorded any bright celestial event and not necessarily the 1054 supernova. Nevertheless, archaeologists have pointed out that when the Moon and Earth return approximately to the same positions they had in 1054, when the Moon is directly over the pointing fingers of the hand in the petroglyph, the supernova was in the position indicated by the starburst symbol.

Meanwhile, the remnant of the 1054 supernova is still visible by telescope today. It is the great Crab Nebula, shown in **16.3**. No longer visible to the unaided eye as it was in 1054, the Crab Nebula nevertheless continues to expand because of the great supernova explosion. Detected ripples of energy still spread out from it like waves of water.

Of the few known early supernova, the Crab Nebula is the most historically well-documented event. Its composition and evolution as a cosmic body precisely dates to the origin of its cataclysmic change. The Chinese recorded it as being as brilliant as a full Moon. It remained visible in the sky during the day for nearly a month. At a distance of 7,000 light-years from Earth, the original star that produced the Crab Nebula was as much as 10 times as massive as our Sun. Because of its historic significance, the Crab Nebula is a favorite target of modern astronomers.

16.3. The Crab Nebula. A six-light-year-wide, expanding remnant of a star's supernova explosion, Japanese and Chinese astronomers recorded this event in 1054, as did, almost certainly, Native Americans. Photo by J. Hester and A. Loll, NASA, ESA.

159

After the 1054 event, observers again recorded supernovae in 1181, 1572, and 1604, the latter two known respectively as Tycho's Supernova and Kepler's Supernova, named for astronomers Tycho Brahe and Johannes Kepler. It is fitting that two of the universe's most spectacular visual events in the last two millennia came in the form of supernovae during the lifetimes of two of the world's greatest astronomers, Brahe and Kepler. Tycho's Supernova made such an impression on astronomers that it was still included on star charts 30 years after the explosion occurred. Figure **16.4** pictures it in a 1603 constellation chart by Johann Bayer, on the upright portion of the chair.

The study of astronomy has been marked throughout history by a number of star atlases. The Bayer atlas was

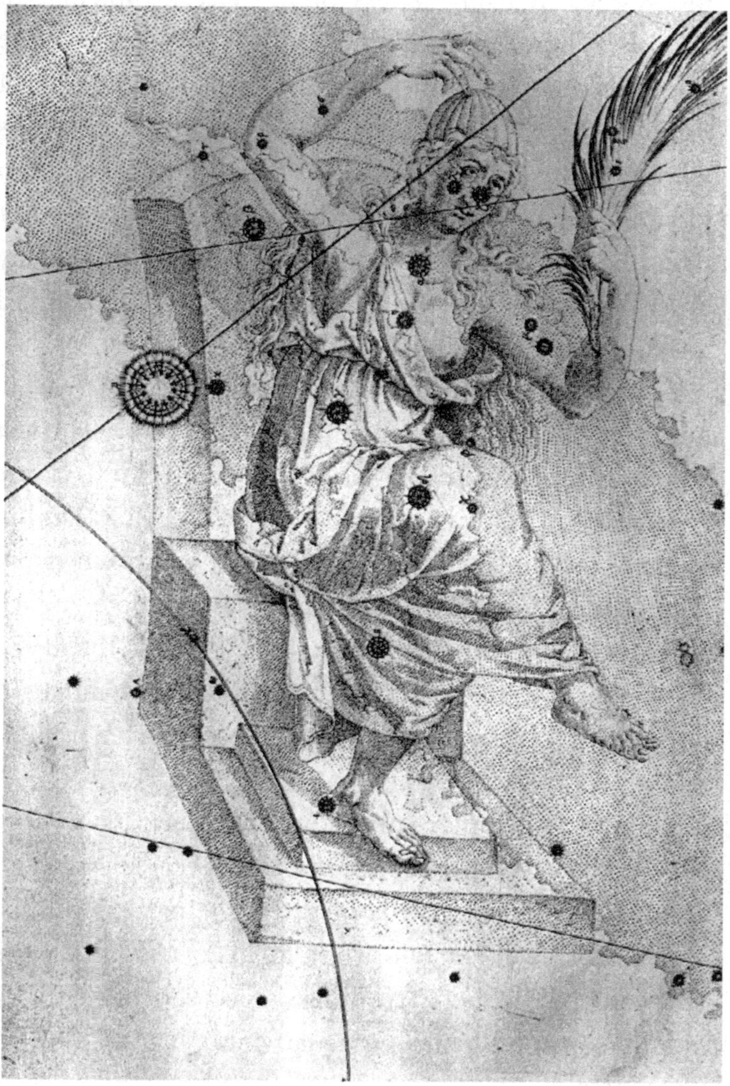

16.4. The Constellation Cassiopeia. Johann Bayer, *Uranometria,* 1603.

the most illustrious of all celestial atlases and the prototype of later ones. It contained 51 star charts produced from engravings, rather than the usual woodcuts of the day. It had a number of distinguishing features, especially its originality of figures, many of which had no known historical prototype. To inform his illustrations, instead of borrowing from the older Ptolemaic tradition, Bayer relied on the more modern view of Tycho Brahe and other distinguished astronomers of the 16th century. Bayer also introduced the system of using letters of the Greek alphabet to designate stars according to their magnitude, the brightest stars being the letter alpha, for example, next brightest beta, and so on. One odd feature of Bayer's work was the reversal of some of the figures. He often pictured them with their backs to the audience, rather than the usual frontal poses. Thus, a star previously described as being on the right shoulder of a figure found itself on the left shoulder.

Supernovae appear on average about every 300 years, but none has been visible for the last 450 years. The last supernova seen from the Milky Way, and the first one observed apparently since 1604, occurred in 1987 in the Large Magellanic Cloud, but it did not quite reach visual magnitude. Pictures taken by the Hubble Space Telescope showed that the aftermath of the 1987 explosion, like the one in 1054, is still very much in progress, with debris expanding at more than ten million kilometers per hour.

Remnants of other supernova have been discovered besides the seven mentioned above. Nevertheless, there are apparently no records of their original explosions. They occurred either before human history or, unlike the 1054 supernova and a few others, observers evidentially did not consider them spectacular enough at the time to be of interest.

Red supergiant stars signal the end of the life of super-hot, massive stars, some known to be 40 times the size of the Sun, and larger. The potential exists for incredible stellar events, when exploding supernovae outshine the distant galaxies in which they occur. These once great stars become small but enormously powerful energy sources, radiating out from the most-distant reaches of the universe.

The death of a mere red giant star results in a dense white dwarf star; the exploding supernova of a red supergiant star leaves a remnant core in the form of a star of even more incredible density.

160

Neutron Stars

The remaining core of a supernova is a small but massive body called a *neutron star*, rotating rapidly, illuminating the remnants of the explosion, and sending out a beacon of energy that may be either visible or invisible. These neutron stars are *pulsars.*

When a red supergiant star explodes as a supernova, it demolishes its outer layer. All that may remain is a dense, super-hot inner core. Instead of a white dwarf, as in the case of a red giant star, this residual core of a supernova is a star of a single nucleus created by the disruption of the star's normal atomic structure. The mass of the nucleus densely packs into an incredibly small area. A teaspoonful of a neutron star would weigh a billion tons. Gravity of a typical neutron star is 100 billion times as great as on Earth.

161

After a star consumes its nuclear fuel, when its self-generating equilibrium is exhausted, it will collapse under its own gravity. The core of the star will shrink rapidly into a mass densely packed by its own gravitational force, until it can shrink no more. In massive giant stars, this collapse of the star causes the outer layers to explode in a violent supernova. Sometimes the entire star shatters, including the core, hurling its remains into space at high velocity to become the material of future star birth. If the star is of the right mass, however, the core likely will survive the supernova explosion to become a neutron star. A neutron star may live another 10 million years in its neutron state.

The genesis of a supernova begins when a massive blue star, its nuclear energy exhausted collapses, contracting to a fraction of its original size. Its matter compacts into a mass of explosive energy a billion times more dense than its original form. The star explodes, and the explosion ejects the star's outer layer deep into interstellar space, causing at the same time a visual intensity that may outshine the galaxy in which it is located.

The filament of the blast of a supernova fades away slowly. For example, the Crab Nebula (**16.3**), now more than 13 light-years in diameter and nearly a thousand years old, continues to expand from its supernova explosion. The expansion of a supernova explosion may last for millions of years. The remnants of millions of supernovae exist throughout the universe, and although several probably occur each year in our own galaxy, not all of them leave a neutron star. Moreover, when they do, they are seldom visible.

162

Little green men

Dubbed pulsars, neutron stars pulsate. They are relatively new to astronomers. First identified by scientists at Cambridge University, pulsars confirmed a long-standing theory for the existence of neutron stars and the supernova process, thus furnishing another one of the important pieces to the puzzle of a star's life.

Statistically, pulsar neutron stars should exist in the millions in the Milky Way Galaxy alone. However, since the discovery of the first pulsar in 1967, the number discovered now runs to fewer than 3,000, including the one shown in **16.5**, seen by telescope as a pulsating pinpoint of light. At about 190,000 light-years away in

16.5. Neutron star in a supernova remnant. Left over from an old supernova explosion, the neutron star is the blue dot at the center of the red nebula. Photo by F. Vogt, NASA, CXC, ESO.

the Small Magellanic Cloud, it is the first neutron star discovered outside the Milky Way.

Typically, neutron stars are ultra-compact stars calculated to be about 30 kilometers in diameter, measuring 1.2 million degrees in temperature, with a magnetic field a trillion times that of Earth. This magnetic field is the alleged source of a pulsar's pulsating radio signals.

Estimates vary on the size that a star must be to become a neutron star, from as low as 1.5 times as massive as the Sun, up to 10 times and more, with the general consensus being about eight solar masses. Astronomers believe that the most massive stars may evolve into something other than neutron stars, while smaller stars transform into white dwarfs. The Sun is obviously much too small ever to become a neutron star.

Most neutron stars, or pulsars, emit strong energy signals, but rarely in the visible range. Only a handful

16.6. Baade's Star. A neutron star located at the center of the Crab Nebula, Baade's Star is spinning at 30 times a second. The innermost ring is about one light-year across. NASA, HST, CXC.

reaches a level of visibility. Most are detectable only by their invisible radio energy. The pulsar in the Crab Nebula, the Chinese supernova of 1054, is a rare exception to this. It is clearly visible and is the closest one to us.

The Crab Nebula pulsar, known as Baade's star (**16.6**), is in the constellation Taurus. The Crab Nebula is the remnant of the great Chinese supernova of 1054.

After nearly a thousand years, its enigmatic presence is as captivating today as when Charles Messier recorded it in the 18th century as M1, the first object entered in his famous catalog of non-stellar objects. Baade's star is at the center of the Crab Nebula. In addition to being the closest to Earth, it is also the youngest pulsar known. Since we know the supernova that produced it occurred in 1054, its age can be precisely dated. This neutron star takes its name from Walter Baade, the German-American astronomer who helped to develop the modern theoretical model of supernovae to describe their explosive behaviors and neutron cores. Baade first recognized that the Crab Nebula is the remnant of the 1054 supernova. Astrophysics remembers Baade for his numerous discoveries, including two classes of Cepheid variable stars whose interpretations doubled Edwin Hubble's original estimates of distances to nearby galaxies, effectively doubling the size of the known universe. In the 1940s, Baade also discovered an opening in the debris shield around the center of the Milky Way galaxy. Through this opening, known today as *Baade's Window*, he was able uniformly to see objects near the center of the galaxy some 25,000 light-years from Earth.

163

Baade's star gained attention as a pulsar in 1968 just a year after the first radio pulsar discovery. Since 1942, the center of attention in the Crab Nebula had focused on suspicion that Baade's star had a connection between it and the supernova that produced the Crab Nebula. Strong radio and X-ray energy beamed across space, and it proved to have a very powerful magnetic field. Nevertheless, it took more than two decades of observation for astrophysicists finally to identify Baade's star as a neutron star.

Baade's star is uniquely interesting because it confirmed the theory that neutron stars are remnants of supernovae and at the same time proved that pulsars and neutron stars are the same.

164

Signals from the first pulsars discovered received the curious moniker of LGM signals, an acronym for Little Green Men. Scientists evidently hoped at first that these radio signals might be coming from other intelligent life in the universe because of their regularity. We know pulsars usually by their radio signals since they are mainly invisible. However, electronic conversion of their radio signals produces visual images that provide pictures of them in some of the universe's farthest regions.

Radio telescopes find invisible pulsars. There are principally two kinds of these telescopes. The Arecibo, Puerto Rico, radio telescope is the world's most sensitive radio telescope. The reflecting dish is 305 meters (1,000 feet) in diameter, 51 meters (167 feet) deep, and occupies an area of approximately 20 acres. The bridge-like construction of the Arecibo employs three-inch bridge cables to suspend its giant aluminum disk above a natural karst sinkhole, located in the Puerto Rican highlands. The Arecibo Observatory is a project of the National Science Foundation.

A second type of radio telescope used to locate pulsars, also developed by the NSF, is the VLA (Very Large Array), or National Radio Astronomy Observatory, situated in the New Mexico desert near Socorro. The VLA telescope comprises 27 separate disk antennae, each one 25 meters in diameter, linked electronically to create a Y-shaped configuration that extends across 36 kilometers. Like the Arecibo Observatory, the VLA is capable of detecting faint invisible radio signals from deep in the universe. While the VLA covers much more area than the Arecibo telescope, its resolution sensitivity is equal to a dish telescope only about half the size of the Arecibo. Meanwhile, the world's largest filled-aperture (i.e., full dish) radio telescope is the Aperture Spherical Radio Telescope completed in 2016 by China.

165

Pulsar signals

Pulsating neutron stars appear to oscillate in predictable cycles. For that reason, for a brief time observers thought them to be like Cepheid stars. Unlike Cepheids, however, which appear actually to vary in

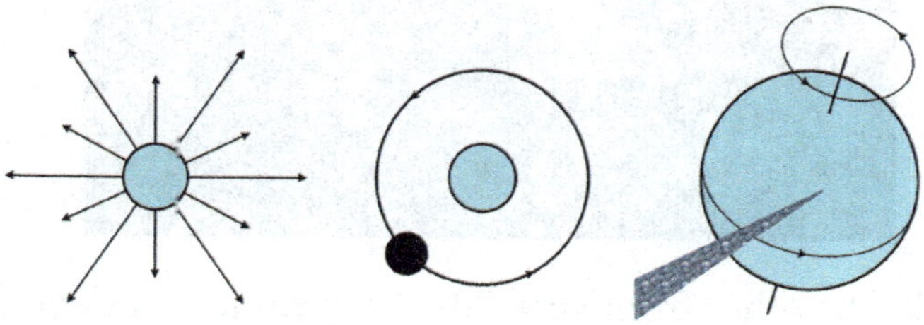

16.7. Pulsar signals. These three models are among several developed by astronomers to explain the regular intermittent signals emitted by pulsars: oscillation (left), orbiting bodies (center), and rotation (right). Rotation appears to be the most plausible solution.

brightness, the oscillation of a neutron star is a steady, intermittent signal likely due not to any change in a star's brightness but probably to its rotation. Figure **16.7** illustrates three possible models of pulsar signals. Oscillation describes the pulsating effect that usually results from the contraction and expansion of an energy source, as happens in the cases of Cepheid and Mira variable star changes. A variety of physical change characteristics may accompany oscillation, such as increasing and decreasing luminance, change of color and temperature, alternating types and levels of radiation, or a combination of these qualities. None of these changes, however, seems to fit pulsar signals.

Intermittent signals may be the result of orbiting bodies. This is another explanation for pulsars. In this scenario, a star's radiation pattern undergoes temporary interruption each time its orbiting companion passes between it and Earth. We see this phenomenon in orbiting binary stars whose orbit planes happen to align with Earth's vantage point. Each time the stars align, their combined visual effect causes the star to appear first brighter and then fainter as one body joins, and then

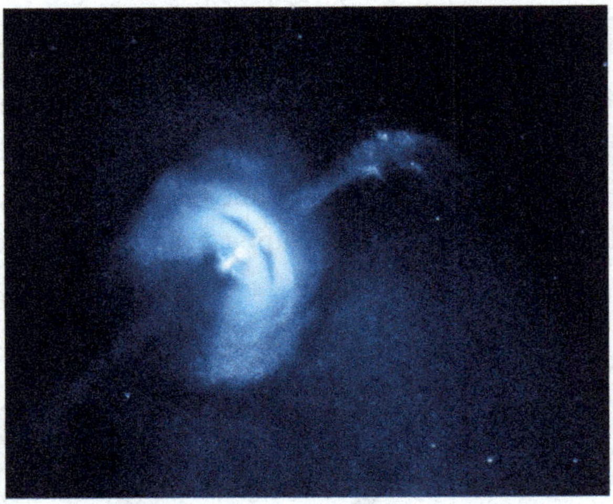

masks the other from view. Most astronomers discount orbiting bodies as an explanation of pulsar signals because of the rapid frequency of pulsars. Their pulsations exceed any known or theorized orbit speeds.

16.8. Vela pulsar. A fast moving jet of charged particles spews out along the pulsar's rotation axis at about 70% of the speed of light. The Vela pulsar is about 1,000 light years from earth, spans about 12 miles in diameter, and makes over 11 complete rotations every second. The supernova that formed the Vela pulsar exploded over 10,000 years ago. Photo by M. Durant, De Martin, NASA, CXC.

What then accounts for the pulsating signals of a neutron star?

Rotation is so far the best explanation of pulsar signals. This assumes a directional energy source emanating from a rotating body. We detect a ray of radiated energy as a pulsating signal each time its rotation passes over Earth. Once thought that the energy beam might theoretically emanate from the space surrounding a neutron star, recent discoveries suggest that the energy signal is probably a function of strong gravitational and magnetic fields of the star itself and their alignment with its magnetic poles (**16.8**).

Still astronomers remain divided on whether the pulsations of a neutron star are due to oscillation or rotation. Most favor the rotation model, even though such bodies would have to be remarkably small in spite of their exceptional mass and would have to be rotating at tremendously high speeds. Neutron stars, if they are rotating, rotate at a rate of four revolutions per second or more. The neutron star at the center of the Crab Nebula is rotating at a rate of 30 revolutions per second. While light traveled from the Moon to Earth in 1.3 seconds, the central star of the Crab Nebula spun around 39 times. Incredible as this is, a more recent discovery of a pulsar in 1982 located one spinning at 642 revolutions per second assuming the rotation model is the correct one for interpreting a pulsar's signal.

166

Exactly how a pulsar works escapes astronomical understanding. Orbiting objects such as other stars that might periodically interrupt the emission of steady signals from a pulsar stand all but ruled out because of the rapid pulsation of some signals. An orbiting body would have to travel at a velocity that defies the laws of conventional physics. Ptolemy once based his argument for a geocentric universe on the expectation that Earth would disintegrate if it required orbiting the Sun at such a high rate of speed. Perhaps conditions exist, yet undiscovered, in which orbits of hundreds of rotations per second are possible.

Likewise, rapid expansion and contraction of a pulsar in the manner of a Cepheid variable are similarly untenable as a cause of neutron star pulsations, because, again, such changes in a body as dense as a neutron star would seem to defy physical possibility. Yet the signals are precise over long periods. Therefore, the rotation model offers the most plausible explanation for the steady, rapid frequency.

167

The lighthouse model

Pulsars are extremely small, perhaps no more than 16 kilometers in diameter. Yet ones mass is much like that of any star; except because of its small size, its mass packs into a density more than a trillion times that of ordinary matter. The collapse of the inner core of a star creates a density of mass that is so greatly concentrated that theoretically, the gravitational and magnetic fields of the star release a single beam of radio energy from a specific area of the star, probably at or near its magnetic pole. This model of energy emission is the lighthouse model. The energy beam of a pulsar emits outward much like light from an airport beacon (**16.9**). As the pulsar rotates, its beacon signal sweeps through space striking whatever happens to be in its directional path, reaching out trillions of miles in a highly concentrated energy signal.

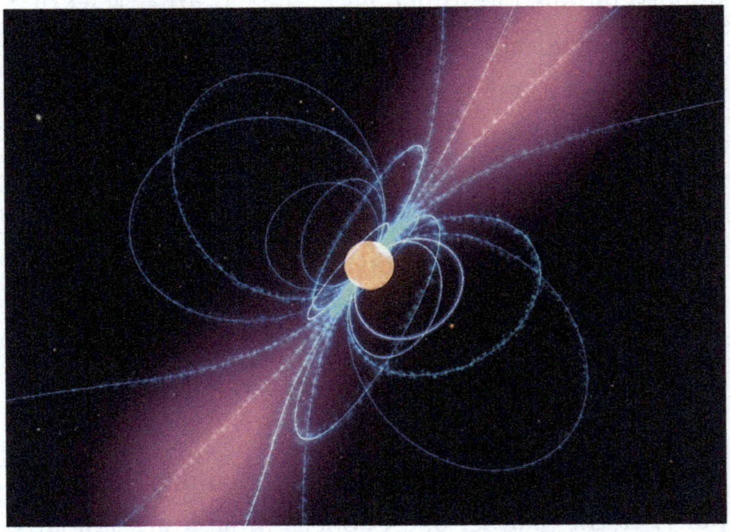

16.9. Lighthouse model. Clouds of charged particles move along the pulsar's magnetic field lines (blue) and create a lighthouse-like beam of gamma rays (purple) in this illustration. This hypothetical model of a neutron star shows the effect believed to account for the pulsating characteristics of neutron stars. A directional beam of energy sweeps through space passing over anything in its path in a way similar to the directional beam of a lighthouse or airport beacon. NASA.

Exactly how the energy beam originates in a pulsar we do not yet fully understand. One theory suggests that a beam of concentrated energy radiates along the magnetic axis of the star. The energy is concentrated as a result of the strong gravitational and magnetic fields, fields that may be so powerful that some energy cannot escape, except by discharge only from the star's poles. The eccentricity of the signal cone, which deviates from the true rotation of the star around its polar axis much like the misalignment of Earth's magnetic pole, results in a wobble that accounts for the intermittent nature of the neutron star's pulsating signal. Although radiation by the pulsar is actually continuous, it appears to be intermittent because of its periodic sweep across Earth. We see it like a revolving lighthouse beacon on the ocean coast would appear as a flashing light each time it rotated into view of a ship at sea.

168

In hundreds of cases, Earth happens to be in the path of a pulsar's beam and is able to intercept its signals. However, this is not always the case. Neutron stars, or pulsars, are likely remnants left over from most if not all supernova events, so neutron stars should be at least detectable in each of the historical examples of supernovae previously cited, and they are not. Baade's star in the Crab Nebula previously mentioned is one that is. However, no pulsars have been detected in either Tycho's or Kepler's supernovas. This may mean that either they do not exist, or the rotations and beam directions of these neutron stars are oriented in such a way that their motions do not align their beams with Earth. There are likely to be hundreds of millions if not billions of pulsars in the universe that for this or whatever reason go undetected.

169

It is possible that the oscillating signals of neutron stars are pulsations unrelated to a star's rotation; that is, they actually turn on and off their signals. Physicists theorize, for example, that the oscillation of a variable star is inversely proportional to the square root of its density; that is, the more rapidly a star oscillates, the greater must be its density. Conversely, denser stars have higher oscillations. If either of these hypotheses is true, even a signal of 642 oscillations per second — clocked for the 1982 pulsar — infers a source that is substantially less massive than the billions of tons per cubic centimeter thought to be associated with a neutron star.

Since the density of a neutron star would theoretically produce an oscillation too great to fit current observations, the rotating beam model again proves to be the best guess of how pulsars work. Pulsars must be rotating because they are too massive to be oscillating. At the same time, the rotation of a body is allegedly also proportional to the density of the body, that is, the compaction of mass affects rotation as well as oscillation. The higher the density is, the higher the velocity of rotation.

Some neutron stars change their rate of radio signals. As mass is redistributed in a neutron star and rearranged in an ever-decreasing volume, the star tends to speed up and slowdown in rotation depending evidentially on the change in its density. Younger pulsars, because they are probably experiencing more redistribution of mass, appear to change speed more frequently. This latter conclusion once more comes primarily from the known age and observed behavior of Baade's star in the Crab Nebula. Astronomers have used this observation to help to confirm the rotation model as the best model for the behavior of all pulsars.

170

Not all neutron stars are the same. A new class of neutron stars called SGRs (Soft Gamma-Ray Repeaters) came to the attention of physicists in 1979 and confirmed in 1986. These powerful magnetic stars give off massive bursts of energy thought to be associated with eruptions on the neutron star's surface due to magnetic stress. Called Magnetars, they may explain why some supernovae appear not to have left a neutron star remnant, because Magnetars, or SGRs, deteriorate rapidly, exhaust their energy, and disappear shortly after exploding as a supernova.

XVIII

BLACK HOLES AND QUASARS

A black hole is a place in the universe where the magnetic field and gravity of mass are so great that no energy of any kind can escape, not even light. Unlike a neutron star with its directed beam of energy, a black hole has no emission of energy and is impossible to "see" by any conventional astronomical means. Figure **17.1** depicts an artist's conception of a

17.1. Black hole. NASA's Hubble Space Telescope revealed a giant black hole in an otherwise normal distant galaxy. M. Helfenbein. Yale University.

black hole. Notice the radio jets radiating above and below the hole. These along with X-rays emitted from the superheated red gas cloud around it are the means by which astronomers detect the presence of black holes. Because of the unknown qualities of black holes, artists' renderings often represent their appearance.

The idea of black holes dates at least to the year 1784 when the British seismologist and astronomer John Mitchell, using Isaac Newton's theory of gravity and the corpuscular theory of light, calculated a theoretical cosmic phenomenon 500 times the size of the Sun that could generate enough gravity to prevent light from escaping. Not long after that in 1796 Pierre-Simon Laplace, a French mathematician, allegedly suggested such a similar phenomenon. Much later, in the 20th century, physicists, who were working at the time with Einstein's relativity equations, "discovered" the first theoretical black holes. (Einstein himself did not believe in black holes.) American John Wheeler in 1968 coined the term *black hole* to describe such a powerful force.

The existence of black holes was under serious consideration by modern astronomers for years; the confirmation of black holes, however, is relatively recent. It was not until 1992 that astrophysicists confirmed that Galaxy NGC 4261 has at its center a black hole (**17.2**). Located relatively nearby at 100 million light-years,

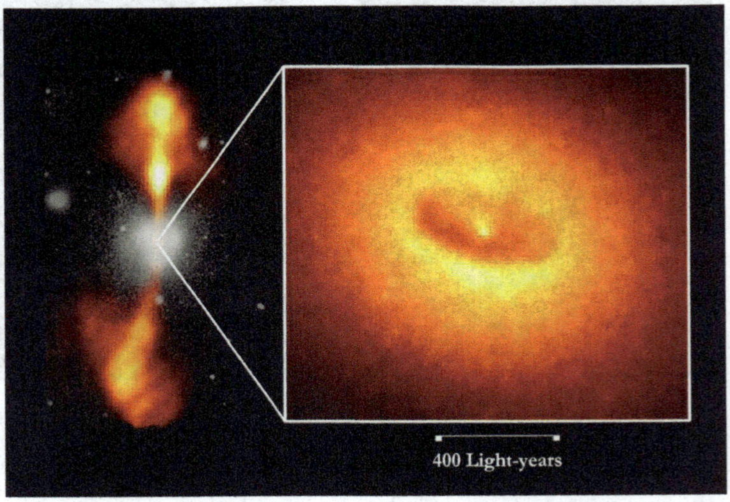

17.2. NGC 4261. NGC 4261 is an elliptical galaxy with a supermassive black hole at its core. The Hubble Space Telescope zoomed into the galaxy center to reveal a doughnut-shaped cloud around an active black hole core. NASA, HST.

toward the constellation Virgo, its core is 1.2 billion times the mass of the Sun squeezed into a space smaller than our solar system and surrounded by a disk of gaseous dust 800 light-years across. Since the discovery of NGC 4261, other black holes have been identified, one of them with a mass three billion times the mass of the Sun.

A black hole represents an entirely different phase in the transformation of matter from that of a star. Black holes are not black dwarfs. Yet, like black dwarfs, black holes are difficult to prove directly to exist because they are invisible to all known conventional methods of detection. They are another form of dark matter. An apt description of a black hole is a cosmic object that has withdrawn from the universe.

172

Event horizon

One explanation of a black hole is that it is the remnant of a cosmic body so massive in its original star or galaxy form that its collapse into a neutron star resulted in a gravitational field so great as to set in motion a state of perpetual collapse. In such a state, gravity constantly increases as more and more matter condenses drawn into the black hole core, compressed there with sufficient density of mass to forever collapse at an ever-increasing rate. The result is that at some point nothing, not even light or radiation of any kind, can escape its gravitational field.

In astronomy, this point of no return is the *event horizon*, the point at which no radiation can escape; all energy collapses, pulled into the collapsing body. An artist's version shows the event horizon in **17.3.** There is no way to see inside the event horizon and nothing that passes inside its boundary can escape.

17.3. Imagined view of an event horizon. An accretion disk surrounds the central core of a black hole. The Event Horizon is the spherical boundary beyond which no radiation can escape due to the gravitational pull of a black hole. Artist's rendering by Newton, ESA, NASA.

Looking at the properties of matter, physicists have worked out a theoretical relationship between the mass of a black hole and the size of its event horizon. For example, a black hole of 10 solar masses would have a diameter of about 60 kilometers. Were it possible for a small body the size of Earth to become a black hole, it would be less than .88 centimeters (½ inch) across, about the size of a marble.

Pulsars and their concentrated directional beacons of energy may be neutron stars that are themselves approaching their event horizons. If so, we can expect them at some point also to withdraw from the universe.

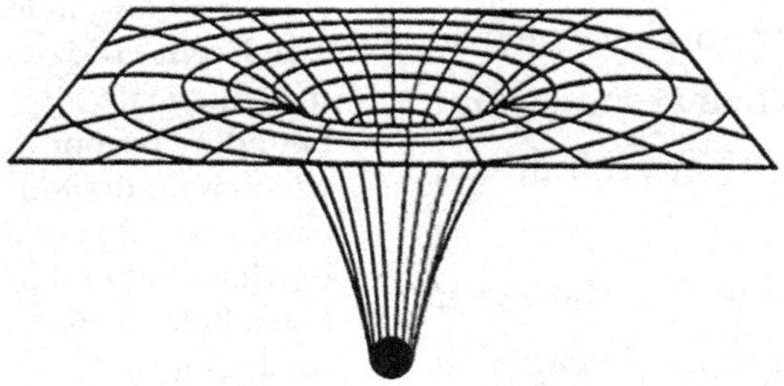

17.4. Gravity bends light. A gravitational singularity lies hidden within a black hole. Northern Arizona University.

Albert Einstein predicted early in the 20th century — and lived to see his theory confirmed — that gravity bends light (**17-4**). It follows then, that very strong gravity of a kind made possible by a massive neutron star can bend light so severely that it actually curves back to its beginning, effectively rendering the source of light invisible to an outside observer because no radiation leaves the source.

How, then, does one "see" a black hole?

As gasses flow into the black hole, they accelerate and rise in temperature. The temperature change causes emission of electromagnetic radiation in the form of X-rays. The measurement of this X-ray radiation is what astronomers use to find black holes. Figure **17.5** illustrates one theory to explain this process. A black hole near another star pulls matter away from that star, accreting the matter to itself. As matter attracts to the black hole, the material accelerates toward the gravitational field squeezed into a superheated state. When the atoms are a few million degrees hot, they emit

X-rays sending out beams of energy just before the matter's final plunge into the black hole.

173

Singularity

If a black hole continues to collapse, it will ultimately reach *singularity*, that is, it becomes as small as it can get — theoretically zero radius. Some astronomers who enjoy testing the boundaries between science and fiction have suggested that numerous black holes of all sizes inhabit space, some existing as mini black holes of pinhead size, but having enormous energy. Some astronomers also have gone so far as to speculate that Earth may actually have collided with one or more of these tiny, high energy specks by citing, as an example, the unexplained Tunguska explosion that leveled hundreds of square kilometers of the Siberian landscape on the Russian frontier in 1908. Mysteriously there was no crater left by that explosion, as might be expected from a meteor impact. No explanation for the event ever received general acceptance. Still, the black hole speck

17.5. The acceleration of mass into a black hole. Matter drawn away from a companion star feeds the smaller but more massive black hole core, traveling faster as it gets closer to the black hole. As the matter accelerates and heats up, it emits x-rays that radiate into space. Eventually the black hole becomes so dense that no energy, including light, can escape. NASA.

theory does not gain much credibility in this case either. It is incongruous that a black hole could have had anything to do with the Siberian event. Incredible mass is required to create a black hole. A star that is massive enough to evolve into a black hole, even a tiny one, would contain enough energy many times over completely to annihilate Earth.

174

Cygnus X-1

Black hole X-rays are irregular, not like the periodic signals of pulsars. This is one of their identifying characteristics. An irregular pattern of X-ray energy, first detected in 1971, appeared to come from the vicinity of a +9 magnitude blue supergiant star in the constellation Cygnus. Dubbed Cygnus X-1, its X-ray signals are bundled in one-hundredth-of-a-second flickers. From these rapid bursts of energy, astronomers calculated that X-1 is smaller in diameter than Earth. An evaluation of the main sequence characteristics of the blue supergiant companion to Cygnus X-1 gives X-1 a mass calculated to be 30 times the mass of the Sun. Doing the math, this black hole candidate then has to have a mass of at least seven solar masses necessary to balance the calculated gravitational relationship between the two cosmic bodies, and to accommodate the accretion phenomenon thought to be active in X-1. An object of seven solar masses cannot be a white dwarf star, and some believed that it was too massive to be a neutron star; therefore, it had to be a black hole. However, it would be another 30 years before X-rays would be associated with black holes.

Astronomers are convinced now that an invisible companion to the blue supergiant in the Cygnus constellation is emitting the X-rays coming from the location of the Cygnus blue supergiant star. The logical deduction is that the invisible energy source is a black hole. Cygnus X-1 therefore is possibly the first black hole discovered. Yet, seven solar masses for a black hole is borderline small. Cygnus X-1 may be a better candidate for a neutron star than a black hole. This ambivalence underscores the problem of trying to predict the behavior of something that is unseen.

Consider, meanwhile, the galaxy M87 in **17.6** as another likely candidate for the location of a black hole. Its large estimated mass and the rapid rotation of a spiral-shaped disk of gas at its core fit the description of a black hole. M87 has a nucleus of matter massed equal to more than three billion of our Suns concentrated into a space no larger than the solar system. M87 is located in the constellation Virgo approximately 50 million light-years away. Its nucleus emits a jet of electrons that scientists think may actually push black holes around space like a self-propelled rocket.

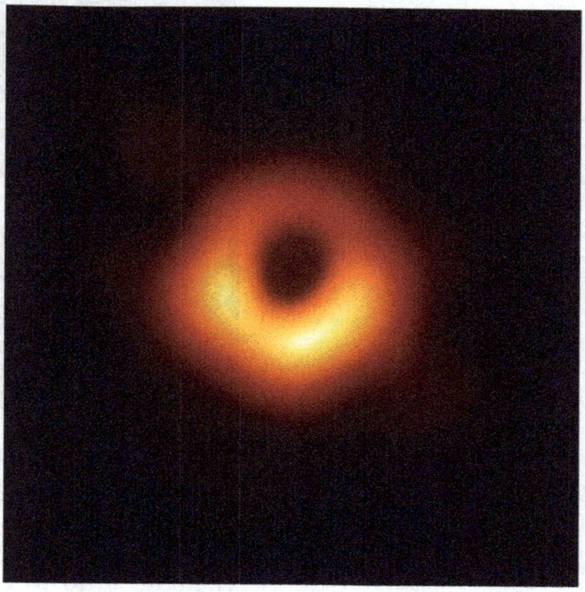

17.6. M87 black hole. The shadow of a black hole is the closest we can come to an image of the black hole itself, a completely dark object from which light cannot escape. The black hole's boundary is approximately 2.5 times smaller than the shadow it casts and measures just under 40 billion kilometers. European Southern Observatory.

As new techniques develop for detecting black holes, evidence is emerging that they may in fact be a very common phenomenon. Theory now points toward every major galaxy having at its center a gigantic energy source — something like a black hole — that will become a black hole as it grows denser and denser. There are theoretically many black holes in our own Milky Way galaxy. The Hubble Space Telescope detected one wandering about in one of the spiral arms of the galaxy, and there appears to be a massive black hole at the center of our galaxy. Something of enormous gravitational force is there that we cannot see. The invisibility of the core of the galaxy generally owes its obscurity to cosmic debris between our vantage point and the galactic center. Nevertheless, other credible theories attribute this invisibility to a black hole, estimated at perhaps as much as 70 million solar masses in scale, with a density of unimaginable proportions. Are we speeding toward an encounter with a black hole at the center of our own galaxy?

17.7. Death Star Galaxy 3C 321. A system of two galaxies rotating around each other, 3C 321 is notable for being the first observed galaxy to smite another galaxy with a blast of energy theorized to be from a supermassive black hole at the center of the galaxy.

Though confirmed only indirectly, astrophysicists accept the existence of black holes. Moreover, some of them may be far more massive than heretofore thought. Galaxies tend to cluster, converging at different points throughout the universe (**17.7**). There are many such galaxy clusters. Great gravitational forces, such as might be exerted by giant black holes, are possibly at the centers of these globular clusters of galaxies. This causes them to be attracted to particular regions of the universe.

One such phenomenon is the Great Attractor, a force so strong that it clearly manifests in the peculiar velocities of surrounding galaxies; yet it is not visible. A picture of the universe taken in the direction of the Great Attractor is included in **17.8.** Toward the constellation Centaurus, this unseen mass is equal to 10 galaxies the size of the Milky Way. It is there, and yet it is invisible. Matter that is visible in that region accounts for barely 10 percent of the gravitational force detected. The galaxies shown in the picture in **17.8** are clustering

17.8. The Great Attractor. Some astronomers believe the Great Attractor is proof of the existence of a major body of unseen Dark Matter in the universe. ESA, Hubble Space Telescope, NASA.

toward this force, which is located approximately in the center of the image. The force is unknown and simply referred to as the Great Attractor. Estimated to be 250 million light-years from Earth, the Great Attractor is so powerful in its gravitational attraction that whole clusters of galaxies, including the Milky Way, are speeding toward it at more than two million kilometers per hour in a stream like a giant river of galaxies flowing toward an unseen galactic ocean.

175

What occurs after a black hole achieves singularity? Is there a point at which total collapse reverses? Does the concentrated matter explode, or does a black hole simply withdraw from the universe to emerge at another place in this or another universe as a white hole?

No one has ever taken white holes very seriously, except perhaps Albert Einstein who thought enough of the idea to join with Nathan Rosen to set forth a model to accommodate a convoluted universe in the Einstein-Rosen Bridge, a theoretical route by which matter could pass from one part of the universe to another. Others, like astronomers Stephen Hawking and Jacob Bekenstein — in their black hole entropy theory — have shown convincing evidence that over time black holes just go away, evaporating as heat and dissipating their great energy into space. Still another theory has black holes combining over time into one gigantic mass, perhaps as prelude to another Big Bang. Under any theory of their dissolution, black holes could well account for the missing matter needed to declare the universe closed.

176

The discovery of quasars

There remain many unexplained things about the universe. Whether the ideas of modern astronomy concerning the birth and death of stars do in fact account for all the major sources of light and energy in the universe, or whether the history of stars and its connection of theories are just so many scientific musings to satisfy the great variety of objects in the sky, remains an open question. The universe is a protean experience, readily taking on different shapes and forms, like the mythological Proteus who could change his shape at will. For just when it would seem that good explanations of the cosmos are at hand, a new class of objects appears, to challenge old ideas and open the universe still further to new inquiry.

This happened not so very long ago with the discovery of a second type of Cepheid variable stars, causing the original estimates of the size of the universe to require adjustment. It has happened periodically throughout the history of astronomy as new technology becomes available. Most recently, it happened with the Hipparcos satellite when new, more accurate measurements of the stars forced revisions of old star distances.

In addition, it happened dramatically in the 1960s with the discovery of a completely new class of cosmic objects in the form of quasars.

177

The brightest objects in the universe are quasars. Just as there are no stars as bright as a galaxy, there are no galaxies as bright as a quasar. Quasars produce more energy than any other known object or collection of objects. A single quasar emits 100 times as much light as an entire galaxy and is hundreds of billions of times as bright as a normal star. Yet a quasar's enormous energy comes from a volume of mass perhaps no larger in diameter than our solar system.

Quasars are associated with galaxies, but they interact with galaxies in different ways. Some quasar galaxies appear disturbed by the gravitational mass of the quasar, while others show little or no effects of quasar interaction.

For years, astronomers thought quasars were just very bright stars, because when seen through telescopes they resemble stars. Then in the 1960s astronomers discovered that quasars are the most-distant objects ever seen, some of them lying at the very edge of the observable universe. Radiation recovered from them originated during the earliest stages of the universe, perhaps as early as when the universe was less than 8 percent of its present age. To lie at such great distances and still be visible, quasars have to be the most energetic sources of radiation ever detected. The precise mechanics of this tremendous output of energy are still unknown.

Quasar is short for a quasi-stellar radio source. Quasars are associated with both light and radio wave energy. Some of them, however, emit neither radio nor light signals. They emit other kinds of higher energy gamma radiation. These are quasi-stellar objects, or

QSO's. Today the term quasar commonly refers to both quasi-stellar radio sources and quasi-stellar gamma objects.

The number of quasars discovered increases with distance from the Milky Way Galaxy, suggesting again that quasars are concentrated toward the edge of the universe, which also assumes that the Milky Way Galaxy is inside the universe and not at its rim. Some quasars exist, in fact, at the very edge of the known universe. These objects appear to be speeding away from us into the unknown at extreme velocities. Quasars are traveling at 20 to 50 percent of the speed of light, with a few predicted to be moving at 80 to 90 percent of the speed of light.

17.9. Quasar 3C 273. The first quasar ever identified resides in a giant elliptical galaxy in the constellation Virgo. Its light takes 2.5 billion years to reach us. Nevertheless, it is still one of the closest quasars. Quasars are capable of emitting hundreds or even thousands of times the entire energy output of a galaxy. One of its jets appears as a cloudy streak, measuring some 200,000 light-years in length. NASA, Hubble Space Telescope.

One theory to explain quasars is that they are the nuclei of old galaxies that temporarily have become exceedingly bright. Others believe that a quasar is an outburst of energy that results when a black hole at the nucleus of a galaxy consumes other stars. If this is the case, a quasar will cease emissions once the black hole has finished feeding until a new source of energy draws into it. This could explain why not all old galaxies are associated with quasars; if there is no black hole, then there can be no quasar. However, according to prevalent theory, most galaxies, including the Milky Way, are collapsing, being pulled into black holes at their centers.

178

The detection of the first quasar dates to 1963. Figure **17.9** is a picture of Quasar 3C 273, the first quasar to be discovered and the one that has the brightest apparent magnitude. Analysists later found images of quasars contained in old photographic plates going back as far as 1887. However, no one at that time thought of them as anything but ordinary stars. We know from recent discoveries that quasars have variable, even pulsar-like behavior. Their brightness changes dramatically over time and in some cases changes sufficiently enough to suggest a possible rotation effect. However, there is no concrete evidence of a pulsar-like rotation in a quasar.

Another characteristic of quasars is their optical jets that may shoot out into space millions of light-years. The jet visible to the left in **17.9** has a complex structure whose radio and optical energies come from very different parts of the jet. One theory about the extraordinary optical brightness of a quasar presumes that we are looking down one of these jets of highly concentrated energy pointed directly at us. The quasar is so bright it obscures the galaxy in which it is located.

179

The first cosmic objects

Quasars, like pulsars, come in varieties both visible and invisible to the unaided eye. We know them mostly by their invisible radio signal and gamma ray brightness. A quasar that is only a faint visible object is 10,000 times brighter in its invisible gamma-ray state. Astronomers have no explanation for its energetic gamma ray outbursts.

While ordinarily invisible, a few quasars are visible by their emitted light. Those that are visible by telescope are mostly high temperature, bluish objects. Much of their light is in the truncated blue and ultraviolet range. Although they emit light in only a small part of the visible spectrum, they may produce light as much as a hundred Milky Way Galaxies, yet be 100,000 times smaller. Visible quasars are relatively small, but their intense energy can cause their brilliance to overpower the radiance of an entire galaxy. In one instance, a quasar that is 1.6 billion light-years from Earth has a brightness much greater than the two dim galaxies next to it.

Visible quasars are generally much closer to us than their higher energy invisible counterparts that are lying at the edge of the universe. This has prompted speculation that quasars, too, like stars, go through an evolutionary process. Scientists think that such highly concentrated energy sources as quasars, especially those located at the edge of the universe, long since have ceased to exist. What we see is how quasars looked in the earliest moments of the universe's beginning. They were probably among the first objects to form after the Big Bang, and almost certainly formed within the first 30 million years that marked the beginning of the galaxies. If the quasars discovered today do still exist, the rate of expansion of the universe has carried them another 10 billion light-years beyond the light detected now. This potentially means that we live in a universe that is not 10-15 billion years old, but 20-30 billion. At such an explosive rate, the universe could double in size about every 10 billion years.

180

The behavior of a quasar is imitative of a collapsed neutron star, only it is millions perhaps even billions of times more energetic. In the early stages of the universe, there may have been cosmic bodies so massive that their ultimate collapse would result in something like a quasar, a galactic scale neutron star. A more fanciful conclusion is that a quasar is the emerging, released energy pattern of a black hole that has somehow traversed the Einstein-Rosen Bridge, to return to a part of the universe different from the one it left. Whatever quasars are they are the enigmas of the universe, defying the models of main sequence star theory, and stirring the brew of cosmic controversy.

181

Meganova and cosmic proportions

Except for its enormous scale, for which there is no comparison, the Big Bang of the universe has many of the characteristics of other exploding cosmic bodies. Could the creation of the universe have been a massive exploding black hole, a meganova? There is among some astronomers a belief that there was already a universe here before the beginning of the one in which we now find ourselves. These astronomers claim to have discovered stars that appear to be older than the galaxies in which they occur, and this, they say, is evidence that something more primal than the material of the Big Bang universe existed for such a phenomenon to occur. Twentieth century discoveries by the ESA satellite, Hipparcos, erased some of the star age discrepancies that previously fueled these contradictions by determining that some stars are younger than originally thought. Nonetheless, diehard theorists continue to argue for a multiple universe origin.

182

The scale of the universe and the galaxies and stars that inhabit it in no way relate to geo-proportions on Earth. It is possible, however, conceptually at least that by elevating ones receptive consciousness to greater dimensions, even than those suggested by the present assumed size of the universe, we could conceive of a universe, or universes, in which our own founding event, the Big Bang creation, was the death and rebirth of a massive cosmic body inside another universe.

Nothing in the universe fits the human concept of ultimate scale. Each new discovery turns out to be beyond imagination. Very early, astronomers combined time and space into a single scale of measurement, the light-year, as tacit admission that the universe is unlike anything in normal human experience.

183

The changing life of a star, despite its sometimes-violent transformation, is a slow process when viewed in Earth time. Recall for a moment The Crab Nebula, an explosive event witnessed more than a thousand years ago in geo-time is still in its explosive motion now a thousand years in the making. There continues a measurable outward movement of the nebula. Similarly, the radial velocities of bodies in the universe and their receding movements away from each other we take to be those caused by the Big Bang. The detected expansion of the universe is a product of the immense size of the continuing explosive action now at least 14 billion years past. The evolutionary cycle of a meganova, as might have created the universe, could well take that many billions of years.

184

Our experience on Earth with explosions is that they are over in a matter of seconds or milliseconds. Because of their relatively small scale, we perceive them all at once within a single time-space reference. This is not so with cosmic events. The larger the scale of an explosion, the longer is the duration of its effects. An atomic explosion on Earth, whose shock wave can be observed spreading outward for several seconds as an Earth-time event, pales in comparison to that of any cosmic event, let alone one sufficiently large enough to begin a universe.

185

What we think of as an instantaneous event on Earth takes on a different scale-duration modality in the cosmos. Even the incredible speed of light cannot overcome the vast distances of space that it must traverse for us to see the results of a cosmic event as one instantaneous occurrence.

17.10. Solar eclipse. Early astronomers and astrologers understood how eclipses occurred. It was nevertheless comforting to the less educated masses to know that the Sun, giver of all life, could disappear and magically reappear seemingly in answer to their prayers. It was a solar eclipse like this one that Aristarchus used in the 3rd century B.C.E. to measure the distance to the Moon, the first accurate calculation of a distance beyond Earth. J. Pasachoff, M. Lu, and C. Malamut.

The origin and demise of a star, for example, are gradual processes. Regardless of its more or less cataclysmic cycle, a star does not suddenly appear and disappear. Stars are so large that neither their formation nor extinction can happen all at once. Moreover, even if it could, we would not see it all at once. Even if our closest star, the Sun, suddenly disappeared, it would not go dark all at once. First, it would be eight minutes before its last light reached Earth, and then it would disappear sequentially. A star is spherical, and part of it is always farther away. The immense size of a star causes light to have to travel further from the outer edges of the sphere than from its closer center. Starting at the center of the solar disk, the extinguishing pattern would travel progressively toward the Sun's outer edge, to linger briefly as a fading, golden ring looking like a solar eclipse before going completely dark (**17.10**). At the same time, Sunlight reflected from the planet Saturn would make that planet visible for three and one-half hours after the Sun disappeared.

We know, of course, that the Sun is not going to go out in this manner. Nevertheless, it is easy to see how early religions imagined such an ending during an eclipse. The ancients believed that the Sun died — went out — and was reborn during a solar eclipse, when the Moon completely hides the face of the solar disc, leaving only a halo of light, as we would experience if the Sun actually did go out. Imagine the fear that we would feel if we thought of an eclipse as our last view of the Sun. It is easy to understand how many ancient cultures placed the Sun frequently at the center of worship.

Multiply the above eight-minute sequence related to the mass of the dying Sun by a mass billions and billions of times greater, as would be associated with a meganova, and the resulting scale of uncertainty for interpreting explosive cosmic events becomes ominous. We would not know a meganova if we were sitting in one.

Jan Muller, after Hendrick Goltzius. ***The First Day: The Separation of Darkness and Light,*** **from** ***The Creation of the World.*** **1589. Engraving, 10 ⅜ in. (26.4 cm.) Harris Brisbane Dick Fund, 1953. The Metropolitan Museum of Art.**

THE EDGE

The history of astronomy is a history of receding horizons.

Edwin P. Hubble.
Realm of the Nebulae, 1936.

Martin Johann Schmidt. *An Oriental Astronomer in His Study*. 18th century.
Pen and brown-black and gray ink, gray wash, 8 ⅛ x 6 5/16 in. (20.6 x 16 cm.)
The Metropolitan Museum of Art.

XIX

EARTH IS AT POINT *A*

186

In the final decades of the 20th century, the extent of the observable universe increased dramatically. Until then, the most-distant object to be measured in the universe was Galaxy 3C 123, a radio galaxy located in the direction of the constellation Cygnus, and so distant that it barely could be observed using a powerful telescope. Its spectrum was possible only by the most sensitive modern electronic devices of the time. This radio galaxy — so called because of its first detection by its radio energy — astronomers, estimated to be eight billion light years distant. Energy detected from this galaxy began its journey eight billion years ago.

Then, in 1998, in an amazing triumph of science the Hubble Space Telescope — one of the most remarkable astronomical instrument ever invented — began to push the frontiers of the universe to astonishing new limits. Astronomers peered into previously unseen regions to see galaxies more than 12 billion light years away. These distant galaxies were among the first comic bodies to form in the beginning of the universe. Yet even greater depths await the new Webb Space Telescope.

187

There are three basic ways of estimating stellar and galactic distances: stellar parallax, the inverse square law applied to magnitude, and spectroscopic analysis. All three methods depend on light or radiation traveling across great voids of space to reach Earth. Parallax measures relatively nearby objects, magnitude is for nearby and intermediate distances, and spectroscopy locates very distant objects. These three techniques —

parallax, magnitude, and spectroscopy combine as one method to calibrate each other where the different ranges of reliability overlap. For example, a nearby object

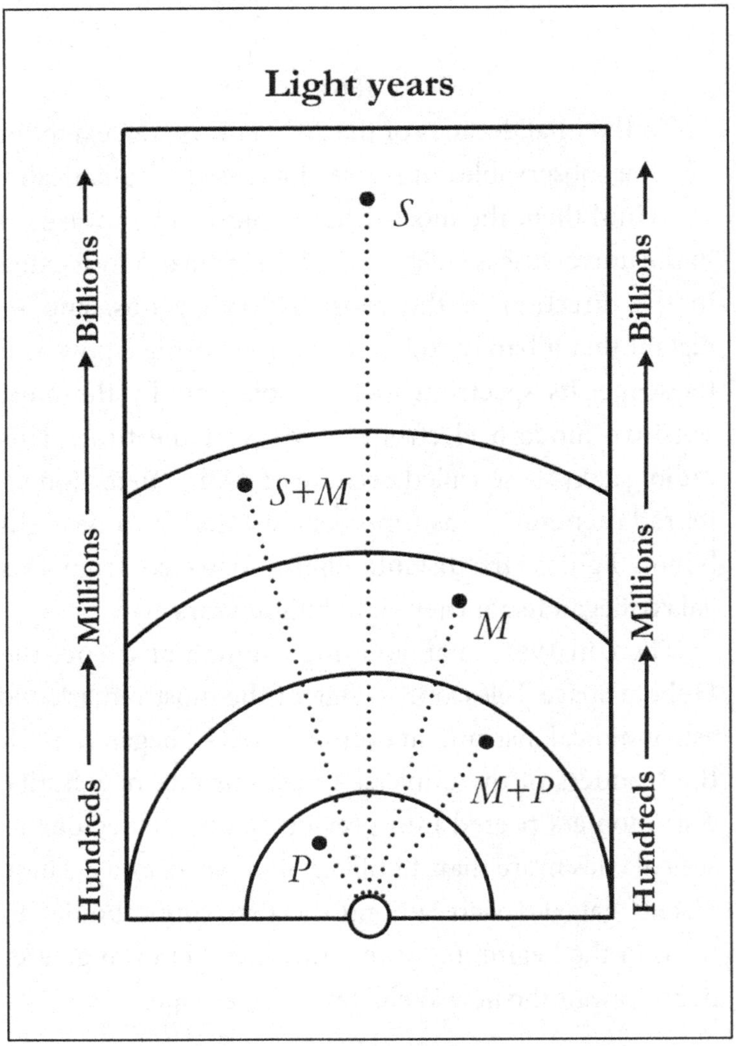

measures best by stellar parallax; farther out, magnitude may prove appropriate (**18.1**). The magnitude of a star whose distance we know by parallax becomes a basis of comparison for applications of magnitude beyond the parallax range. Similarly, if magnitude reliably predicts the distance of an object, then its spectral analysis will infer distances to like objects outside the confidence range of both parallax and magnitude. Objects within a

18.1. Cosmic distances measured by Parallax (P), Magnitude (M), and Spectrography (S).

range of overlapping techniques obtain confirmation as to their predicted distance by more than one method. In this way, it is possible to go from star, to galaxy, and beyond like stepping-stones to the edge of the universe. Looking at the diagram in **18.1**, stellar measurement by parallax *P* combines with magnitude *M*+*P* to confirm distances to relatively nearby stars. Magnitude *M* combines with spectrum measurement *S*+*M* for stars and galaxy distances. Spectral analysis *S* is the measurement technique for distance estimation to the deepest reaches of the universe. Parallax *P* is dependable traditionally up to 100 light years; magnitude *M* is acceptable in the millions of light years; and spectral analysis *S* predicts distances in the billions of light years.

188

Stellar parallax, magnitude, and the inverse square law

The parallax method is a geographical technique adapted to planetary and stellar measurement. On Earth, when we know the distance between two points, the distance to a third point calculates from angular measurements and the application of trigonometry. This parallax surveying technique we know works on Earth because verification of terrestrial results by direct measurement are possible in ways that do not rely on trigonometry. Since the parallax method works on Earth, the assumption is that it will be applicable to the solar system and nearby stars, even though confirmation by direct measurement obviously is not possible.

The sphere of confidence for the parallax method extends out anywhere from 100 to 300 light years (946 trillion kilometers), depending on the location of the star and the observer's vantage points. The application of an alternative technique of measurement — magnitude, for example — to the same objects within the parallax sphere of confidence makes it feasible to overlap the two technologies and predict distance measurement beyond the limitations of the parallax method alone. This assumes, of course, that the new technology works the same way and equally well, as it did within the parallax range, when extended to a larger universe beyond the parallax confidence sphere.

189

If we know the luminosity of a star, we calculate its distance from its visual magnitude. Brightness is inversely proportional to the square of the distance from the viewer to the light source. The problem with the inverse square technique of measurement is that both the magnitude and luminosity of a star are necessary for the technique to work. Luminosity is the light energy emitted or reflected by an object at its surface. However, there is no way to measure directly the luminosity of a star at its surface. Without this information, it is impossible to know whether a star is bright because it is nearby or because it is especially luminous. This question requires an answer by a means other than direct measurement.

190

Visual or apparent magnitude is the perceived brightness of a luminous object from Earth. Apparent magnitude easily ensues simply by looking at a star and gauging its brightness. The system today, of course, is much more technical than trusting the human sense of sight. At first, applications of the magnitude method of distance estimation assumed merely that all stars were exactly like the Sun. We know the visual magnitude of the Sun. Furthermore, we know its distance through the parallax technique; so, it is a simple matter to calculate its luminosity. Astronomers accurately measured distances to the Sun and the planets plus the distances to some of the nearest stars many years ago by the stellar parallax method.

191

Overlapping results

Absolute magnitude is the brightness of a star at a standard distance of 10 parsecs (310 trillion kilometers). Absolute magnitude says, in the example of the Sun, how bright the Sun would appear at 10 parsecs, or 32.6 light-years instead of the 150 million kilometers distance from which we see it viewed from Earth. It is possible, in the case of the Sun, to calculate its absolute magnitude by applying the inverse square law to its observed, or apparent, magnitude. From the greater distance of 10 parsecs, the Sun's absolute magnitude turns out to be +4.8. This is how bright the Sun would appear at 10 parsecs. A star at this magnitude is barely visible to the naked eye.

If one wanted to experience what the Sun would look like from outer space at its absolute magnitude distance, the Sun compares favorably with the star Chara in the modern constellation of Canes Venatici, the Hunting Dogs. Visible in the northern skies during spring 27 light years distant, Chara is a yellow-orange star about the same size and temperature as the Sun. Because Chara rotates at a speed similar to the Sun and like the Sun is a lone star, unlike many stars that appear in binary form or in clusters, astronomers mention Chara as a candidate for a planetary system. Spectroscopic analysis has detected an unseen companion in orbit around Chara. In any event, Chara provides a good opportunity to visualize how the Sun might appear when viewed from approximately the same distance. Meanwhile, anyone or anything observing us from a planet in the Chara system sees us about the same way we see them from Earth.

Based on the magnitudes of the Sun and the star Chara, it might be inferred that any star of similar magnitude, +4.8, lies approximately 10 parsecs away, assuming that it is a star like our Sun. This, of course, is a big assumption. We know that the Sun is typical of most stars. If other stars were of equal luminosity to the Sun, calculation of distance based on their relative magnitudes would be an easy matter; easy but potentially inaccurate, because not all stars are of equal luminosity and not like the Sun.

If the Sun or similar star is beyond 70 light years from the viewer, it is invisible to the unaided eye because apparent magnitude drops below the +6 magnitude visual threshold.

At absolute magnitude of +4.8, the Sun is 10 parsecs distant, or about 32.6 light years. Therefore, any dim star of approximately +5 magnitude, if it is like our Sun, lies with a high degree of confidence at about the same distance of around 30 light years. At the same time, a star like Pollux in the constellation Gemini, to use another example, is at about that distance as Chara. Pollux is 31 light years away. At magnitude +1.2, it appears much brighter than Chara. Therefore, Pollux cannot be Sun-like because of its higher absolute magnitude. It must be a hotter, more luminous star than the Sun. Pollux is, in fact, 32 times more luminous than the Sun.

192

If a star is nearby — within 200 parsecs — the inverse square-magnitude estimate is subject to double-checking by the parallax method. In this way, it is determined if a star is indeed comparable to the Sun, or if it has important differences. If the parallax and magnitude measurements produce different distance predictions, the star becomes the subject of closer study to see just how it differs from the Sun.

The precision of the inverse square-magnitude technique depends on the known characteristics of the star. Today it is possible to learn about the nature of a star and to know if it is or is not similar in size and composition to our Sun.

193

Stellar parallax has a high degree of certainty when measuring relatively nearby objects like planets and neighboring stars. However, parallax is not reliable applied to more distant objects such as galaxies and far away stars because the angles of measurement to those objects are too small.

Spectroscopic parallax

Meanwhile, beyond the overlapping range of parallax and magnitude, confidence in magnitude distance estimates begins to drop, also. Magnitude provides approximations of distance for star and galaxy magnitudes to about magnitude +23, providing we know the type of star or galaxy. Anything dimmer than magnitude +23 is generally below the ability of current instrumentation to measure reliable photometric brightness. Magnitude +23 is approximately the range of the strongest telescopes. Although observations of low magnitudes are possible, the confidence level for magnitude as a distance measurement tool begins to drop at about magnitude +16.

How, then, are distance estimates possible when distances are known to be millions of times greater than anything determined by the parallax or magnitude methods? Here 20th century astronomy enters.

194

Max Planck's black body theory as applied to stars was the first leap of achievement into greater cosmic distances. Up to that time, the only tools available were parallax and magnitude, but magnitude depended on knowledge of the star undergoing measurement. The discovery that a star's magnitude directly relates to its color temperature provided a major breakthrough. It penetrated the known celestial ceiling to reach as far as the source of reliably detected light.

The Hertzsprung-Russell Main Sequence theory improved the color temperature benchmark when astronomers discovered that the vast majority of stars appear to be very predictable in their color-magnitude relationships. Spectrographic analysis of a star provided many things, among them a star's color temperature. Color, in turn, predicted a star's luminosity, from which may be calculated its absolute magnitude. From here, it was simply a matter of calculating the distance between absolute and observed magnitudes using the inverse square law. This method is spectroscopic parallax. That is to say, spectral star type gives absolute magnitude, and when combined with apparent magnitude it gives distance.

195

Star analysists soon developed important calibration techniques to improve on the spectroscopic color temperature method. Cepheid variables, for example, proved to be invaluable in this respect. Cepheid variables are stars whose periodic changes in brightness correspond to their luminosity. Unfortunately, there are no Cepheids within the reliability range of the parallax method; so, total unerring accuracy in the application of Cepheids is not possible. Still, it is possible from spectrographic analysis of stars that are near known Cepheids to establish a very good scale of distance related to the period-luminosity function of a Cepheid star.

Over the years, astronomers refined Cepheid calibration to the point that when the period of a Cepheid star is known a good amount of confidence exists in setting its absolute magnitude, just from the frequency of the star's oscillation. Moreover, when absolute magnitude is known, inverse square calculation may be applied and distance computed for a Cepheid by working backward from its observed magnitude on Earth to its absolute magnitude as determined by its period-luminosity function.

196

Open star clusters and stellar motion

From the vantage point of Earth, stars appear to the unaided eye to remain stationary in the sky. Just as an airliner traversing the sky overhead appears to move slowly relative to our place on Earth, stars are at distances so great as to render their movements undetectable in normal human experience. Most stars appear to the casual viewer to remain stationary, but in reality, this is not the case, and astronomers are able to see even the slightest movements in stellar motion.

Objects that are parallel to our line of sight, that is, objects that are advancing or receding radially away from or toward us, are very difficult to detect in their movements at any distance, but particularly so if they are far away. Objects moving perpendicular to the line of vision, on the other hand, are easier to judge for their motion. In astronomy, this perpendicular motion across the plane of vision is *proper motion.* Proper motion is the term for lateral motion as opposed to receding or advancing parallel motion, which is *radial motion.* As one example, the star that has the greatest proper motion from our vantage point is Barnard's star, located in the northernmost part of Ophiuchus, the constellation of the Serpent Holder. In 200, years Barnard's star moves a little more than the perceived diameter of the Moon, or about 12 inches. It has moved approximately two meters in our plane of vision in the 1,800 years that have elapsed since Ptolemy's time. Barnard's star is too dim to be visible now, but by the year 11,800, it will have traveled from a distance of six light years to only 3.8 light years to be our closest star neighbor. Today it is our second closest neighbor behind Alpha Centauri.

197

An open star cluster is a group of stars of about the same age, moving through space together in a common motion. Some of these clusters, like the Hyades and the

Ursa Major (Big Dipper) clusters are relatively close to the solar system. Because they are nearby, astronomers can determine the intrinsic brightness of the different star types. They begin by observing the motions of the stars in a cluster. Stars in an open cluster are moving as a group. However, their individual proper motions when viewed from Earth, point to a convergence at a common point. By studying old star photographs, and by other methods, analysists deduce this point of convergence mathematically. From the resulting geometry of moving clusters, distance is then calculated. Nearby clusters, such as the Hyades, also provide for the application of parallax measurement. By applying different techniques of measurement to the same star cluster, a reasonably accurate scale of distance evolves. The cluster then becomes a veritable laboratory for the examination of star compositions and behaviors. Because distance is known for stars in a cluster, idiosyncrasies related to the magnitude of a star, such as its composition or the period-luminosity cycle of a Cepheid star, may be precisely related to a distance-magnitude scale and used as distance calibration for other clusters or for nearby galaxies. These data generate a scale created to correlate with the main sequence H-R curve of different magnitudes associated with various kinds of stars in the cluster. Application of this tool makes possible the calculation of distance. For instance, the Hyades cluster, a cluster of about 100 stars in the constellation Taurus, has played a key role in distance measurement. The angular changes of stars in the cluster as they move away from Earth mathematically translate into distance measurement. This information supports many other distance predictions in the universe. Lying at a distance of 150 light years, the Hyades has been estimated to be about 660 million years old, relatively young in cosmic time. Since all the stars in a cluster are assumed to be about the same age, important differences in star types can then be studied based on a constant age factor.

Meanwhile, the Pleiades, another open star cluster, measures by satellite parallax to be 380 light years away, and only about 100 million years old. The Pleiades cluster has been of particular interest to astronomers because of its interesting composition of many star types and nebulae, including several exotic-type brown dwarf stars that were previously only hypothetical objects.

The closest star group to the solar system to qualify as a star cluster is the constellation Ursa Major. We estimate Ursa Major to be about 75 light years away and about the same age as the Hyades. It has been determined that the stars in this familiar Big Dipper star pattern share a common proper motion. Although they are at very different radial distances from Earth, they appear to be moving through space as a common body toward us, pointed in the direction of a convergence point in the constellation Sagittarius, toward the center of the Milky Way Galaxy. Compared to the Hyades and Pleiades, however, Ursa Major covers an enormous section of the nighttime sky, spread over a volume 18 by 30 light years in size.

198

Each distance prediction technique complements another, as ways to reach deeper and deeper into space. To give yet another idea of how this stepladder effect works, consider a closer examination of Cepheid variable stars when associated with an open cluster. There are unfortunately no Cepheids in nearby clusters. There are, however, Cepheids located in remote clusters whose distances are known from comparisons with star characteristics in nearby clusters; that is, by comparing clusters of like characteristics, distances are extrapolated and extended until a cluster is found that does contain a Cepheid variable star. That Cepheid we then calibrate to the known distance of its parent cluster. Consequently, Cepheids of similar behavior when found in distant star clusters and in galaxies even further out can tell us the distances to these objects.

When Cepheid variables appear in clusters of stars, we assume that stars in that cluster are all at about the same distance. This is how Henrietta Leavitt made her famous discoveries about Cepheids in the first place. When the variable of distance is controlled, different magnitudes of stars must be differences owed to factors other than distance, such as luminosity. Closer study of the characteristics of several stars in a cluster can provide clues needed to adjust further the distance prediction process, particularly when such stars are in motion together. Each new discovery permits further enlargement of the sphere of confidence about cosmic distances and often changes the focus of astronomy. When Edwin Hubble discovered a Cepheid star in another galaxy, it completely altered the understanding of the scale of the universe. Cepheid variables, initially good for distance measurements up to about 30 million light years, greatly expanded when the Hubble Space Telescope nearly doubled their distance before very distant Cepheids became too dim to be useful. After that, estimation techniques passed from individual stars and star clusters in galaxies to whole galaxies themselves.

XX

INTO THE ABYSS

199

Stars and galaxies travel outward at great speeds in a continuously expanding universe. Objects in the universe are moving radially away from each other at speeds that are proportional to their distances. Those furthest out are moving the fastest. Expansion appears to be outward in all directions from our own Milky Way galaxy. However, this is a relative interpretation because the same observation would be true no matter where in the universe the viewer happened to be. It is like seeing the effects of an explosion from the inside. Nearby objects are receding relative to the Sun and Earth at between 100 and 500 kilometers per second. Farther out, the galaxies are moving at ten times that speed or more. Our own galaxy has an absolute velocity of between 600 and 670 kilometers per second. Absolute velocity is velocity measured relative to the primeval background radiation of the universe, known in astronomy as the *standard of rest* of the universe.

200

Doppler Effect

Emitted wave fronts travel outward uniformly in all directions from a radiating source. A source at rest or moving at a relatively slow speed produces the wave pattern shown in **19.1 *a***. If, however, the wave source is

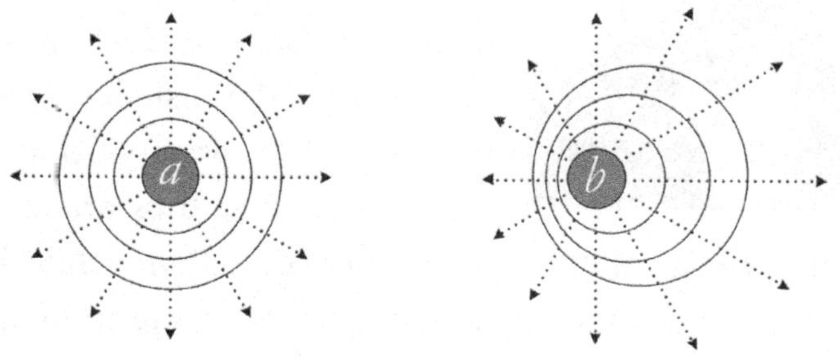

19.1. Wave fronts for a stationary source *a* and a moving source *b* demonstrating the Doppler Effect.

in motion, particularly if it is moving very fast, the wave pattern will look more like **19.1** ***b***. The wave fronts compress in the direction of the movement and expand in the trailing direction of the radiating source. This is the Doppler Effect.

A stationary energy source will emit energy of equal wavelength in all directions. When applied to sound wave energy, an observer at any point in the proximity of the energy source will hear a uniform sound, providing that both the sound emitter and the observer are stationary. If, however, either the energy source or the observer is in motion relative to each other, as illustrated in **19.1**, the waves will be compressed or expanded depending on the vantage point of the observer, that is, whether the motion is toward or away from the observer. A normal sound coming from an approaching source will be compressed and will sound at a higher pitch; while the same source moving away from the observer will cause the frequency of the waves to be spread out, thus sounding lower in pitch.

201

The name Doppler Effect is for Johann Christian Doppler, the Austrian physicist who first applied it to sound waves in 1842 when he loaded a brass band onto a moving railway car to prove his point. A more familiar but related example is the whistle of a train. The whistle of a moving train will have a higher pitched sound in the approaching direction than it does receding away from the listener. This is because the wave fronts are compressed, and thus arrive more frequently in the approaching direction, while being spread out to arrive less frequently as the train speeds

19.2. Portrait of Johann Christian Doppler. Unknown photographer, bef. 1853.

away. Doppler (pictured in **19.2**) showed that sound waves would stretch or compress according to the direction and speed of the object that emits the waves relative to a fixed point. The noticeable change in the sound of the train and its whistle from a sustained pitch as it approaches, to a tremulous, lowering tone as it passes and moves away is due to the speed of movement of the train relative to the fixed position of the listener. The same effect would occur if the observer were moving relative to a stationary train.

The Doppler Effect is responsible for other sound phenomena such as the sonic boom of high-speed aircraft.

The Doppler Effect also applies to light. Light waves travel in a manner analogous to sound waves, except at tremendously higher speeds. Instead of changes of pitch in sound, however, altered light waves produce changes in color.

When applying the Doppler Effect to sound, it is a simple matter to compare the moving object to its at-rest, normal sound, or, in the example of a moving train, to hear all three states in progression: approaching, normal, and receding. This is in effect what happens at a rail crossing when the sound dramatically changes as the train moves through the crossing from one direction to the other.

The application of the Doppler Effect to stellar and other cosmic movements, however, is less convenient than its application to a moving train. There is no relative vantage point from which to view the Doppler transition in light; therefore, there is no visual way to know what is normal in a star's color and whether it changes due to the Doppler Effect. We need a way to establish its normal, at-rest color. We do this by spectroscopic analysis and the creation of unique spectral line patterns.

201

Spectroscopic analysis

The basic physical elements that comprise all matter in the universe produce unique energy patterns that appear as lines seen distributed over the electromagnetic energy spectrum. For example, the element Hydrogen — the major element in stars — has a distinct spectral

pattern on the visual spectrum. Helium and Carbon, meanwhile, two other elements commonly found in stars, show uniquely different line patterns that distinguish them from each other and from the simpler Hydrogen element. Spectrograms, or photographic prints, of these three elements are in **19.3**. Spectrograms derive from the famous Fraunhofer Lines. The basis for the science of spectroscopy was set in 1814 when the German physicist Joseph von Fraunhofer discovered lines in the solar spectrum. The spectrograph, though highly sophisticated as a modern instrument, is the same in concept little changed since Fraunhofer's time.

19.3. Spectrographic prints, or spectrograms, of three elements: hydrogen, helium, and carbon.

In normal elemental states, spectrographic lines always appear in the same line pattern and in the same position on the spectrum. Therefore, once identified, the line code of an element distinguishes it from other elements with or without its inherent color associations. While stars are predominantly made of hydrogen, they are nonetheless compound bodies made up of multiple elements. So the light arriving from them may show by spectroscopic analysis a multitude of element line codes, all overlapping each other as one continuous spectrum of lines. Experts in spectroscopy, with the help of modern technology, know how to decipher and read these very complex code groupings.

Isaac Newton showed that light is composed of a spectrum of colors. Therefore, to determine the baseline spectral definition of any element, such as hydrogen, for example, we need its color spectrum. To obtain the spectrum of an element, we burn or heat it to its luminous state. The emitted light, passed through a prism, as in the original Newtonian experiment, produces the element's spectrum.

An instrument called a spectrograph measures the sample of light across its different wavelengths or frequencies. A photoelectric cell detects the relative amount of energy at each frequency and an apparatus draws a graph of the light's composition.

The basic setup for a spectrograph is in **19.4**. The spectrograph is composed of a slit through which the light passes; a series of collimating lenses to make sure the light is coherent; and a prism or grating to refract the light into its unique elemental spectrum. An electronic sensor or recording device placed in the path of the spectrum image records the relative energy in each of the spectral bands. When the recording medium is film, the resulting photographic record is a spectrogram.

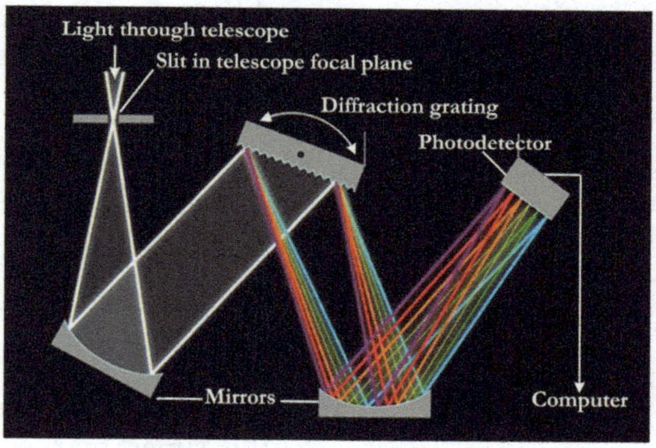

19.4. Spectrograph configuration. George Retseck, *American Scientific Magazine*

Pictured in **19.5** is a spectrogram of the solar spectrum that looks much like Fraunhofer's earliest spectrographic results. The solar spectrum produces a pattern of dark lines distributed across the spectrum.

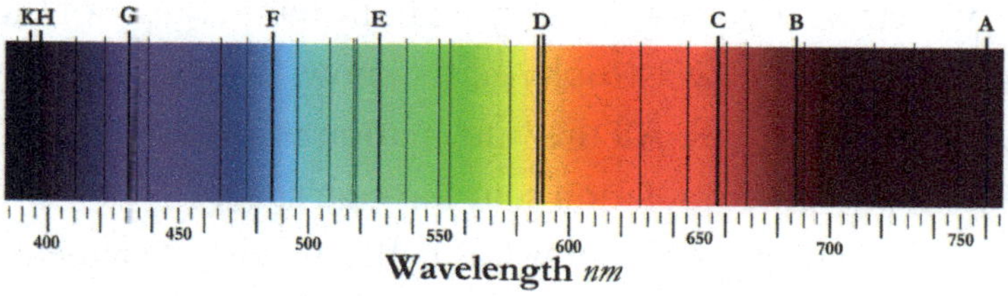

19.5. Solar spectrum. A graphic representation of Fraunhofer Lines overlaid on the electromagnetic spectrum replicates the 1814 discovery by the German physicists of the absorption elements of the solar atmosphere.

Fraunhofer systematically studied and measured the wavelengths of these lines and gave them letters of identification. Years later, scientists noticed that Fraunhofer's map of dark lines coincided with emission lines identified in the spectra of heated elements. They deduced that the dark lines of the solar spectrum were the result of absorption by chemical elements in the solar atmosphere. The discovery opened the door to spectrographic analysis of the makeup of cosmic radiant energy emitters. Modern technology can record spectral emissions in the ultraviolet, infrared, and other portions of the electromagnetic spectrum, as well as the traditional visual spectrum.

202

Statistical parallax

Earth and Sun are of the same elements, only in different proportions and under different conditions of temperature and density. The basic chemical elements — Hydrogen, Helium, Carbon, etc., alone or in various molecular combinations, account for all matter. Each element of matter has its own unique spectrographic code. These element codes show up as lines on a spectrogram, a black and white print of the spectral code similar in appearance to the bar code printed on packaging used in product identification and pricing in the super market. The strength of each element will appear in the spectrographic code proportional to the presence of the element in the star or galaxy undergoing analysis. If the star is composed mostly of hydrogen, then the hydrogen lines in the code will be stronger than the lines of other elements. The strength of the lines in terms of their widths may also vary with the temperature of the star. This is how astrophysicists know the composition of cosmic bodies. All the information is available in the light and other radiation sent through space.

Knowledge of the physical behavior of the elements composing the stars, and how these patterns behave in spectrum analysis, continues to be the most significant advancement made toward the understanding of the composition of the universe. Consider next how this knowledge applies to the Doppler Effect.

203

When a light source is moving fast enough relative to the viewer, the Doppler Effect will occur, in this case compressing or expanding the light waves. In sound, the Doppler Effect manifests in the auditory changes in the sound waves. In light, the result is a change in color. An object speeding toward the viewer at very high speed will appear bluer than the same object would at rest, while one traveling away at high velocity appears redder. Short wave lengths characterize the blue portion of the color

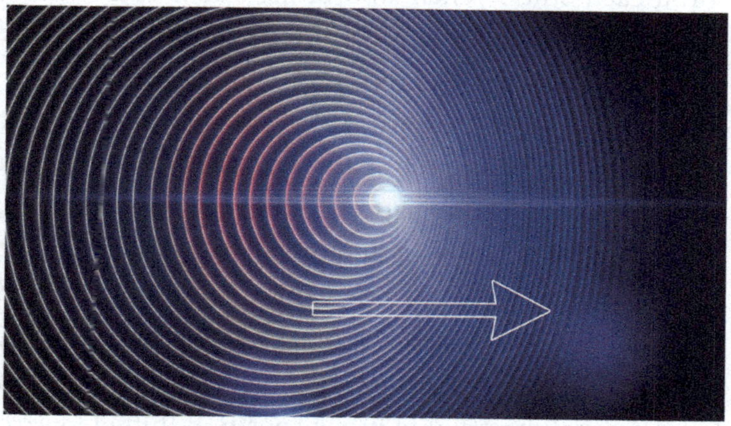

19.6. Red shift-blue shift Doppler Effect. A monochromatic light source moving at high velocity toward the viewer will appear blue while it will appear red moving away from the observer. L. Calçada, ESA, Hubble space telescope.

spectrum, while the red portion has longer wavelengths. This phenomenon in astronomy is the *blueshift* and *redshift*. Usually the objects under analysis do not appear bluer or redder to the eye because the shifts are relatively small in terms of their visual effects. Nevertheless, they are measurable to a high degree of significance in spectroscopic analysis (**19.6**).

Light emitted from a moving source will appear as redder or bluer depending on whether the source is moving toward or away from the viewer. The unique spectral pattern of each element does not change but adjusts on the color spectrum according to the Doppler

Effect toward the blue or red depending on which direction the light-emitting body is moving. Toward the viewer, the wavelengths compress into the shorter wavelength, blue end of the spectrum; away from the viewer, just the opposite occurs; the wavelengths lengthen toward the red end of the spectrum. The blueshift or redshift will indicate whether the light source is approaching or receding and at what velocity. Astronomers have confirmed the accuracy of the spectral wave shift technique by observing binary stars, stars that orbit each other. When an orbiting star is approaching in its orbit, its light is blueshifted; when it is moving away, the same light is redshifted.

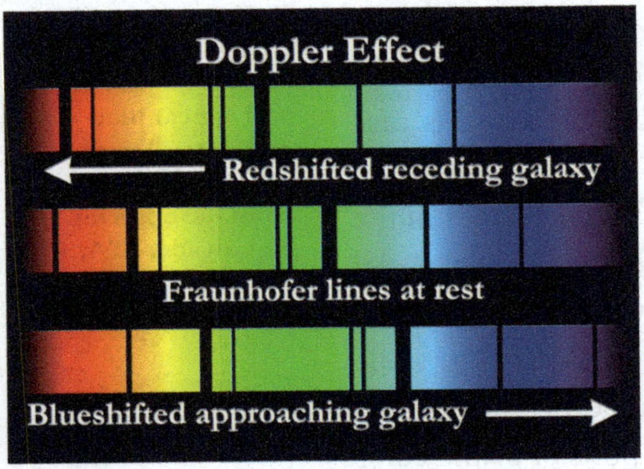

19.7. Doppler Effect shift. The spectral lines of a body at rest will shift toward the blue or red parts of the spectrum depending on whether the body is moving toward or away from the observer. West Texas A & M University.

The reference patterns of element spectral lines, or spectrograms, from deep space remain unchanged from how the same element patterns appear in the laboratory, regardless of a wavelength shift; that is, the complete pattern for each unique element surveyed shifts as a whole. Thus, it is possible to see the displacement of the spectrographic pattern from its normal at-rest position on the visual spectrum to its new position shifted toward red or blue, as illustrated in **19.7**. Scientists are able to read very small shifts in these patterns and to translate the amount of the shift into the relative speed of the light emission source. Fully automated redshift analysis today makes very fine readings possible.

204

The faster an object is moving away from Earth the greater will be the redshift. The greater the velocity of the object, the more will be the spectral displacement. The percent of shift seen in the spectrum is the percent of the speed of light at which the body is moving. For example, a shift of two tenths of 1 percent (.002) is 600 kilometers per second (300,000 km/sec, multiplied by .002=600). Most velocities calculated by redshift analysis are much faster than 600 kilometers per second. The fastest receding star so far detected in our galaxy exhibits a redshift equal to 20 percent of the speed of light, or 60,000 kilometers per second. Beyond our galaxy, however, objects have been detected which appear to have a velocity above three-quarters of the speed of light.

205

Hubble's law

Much of the knowledge about the movement of stellar objects owes to the work of Edwin Hubble. Applying the Doppler theory to light waves, Hubble was able to calculate mathematically a great deal of information about distant celestial bodies from the evaluation of light from these bodies. He discovered in 1929, for example, that the farther away an object is in the universe, the greater its redshift, and the faster it appears to be moving away from our solar system. This relationship of redshift to speed and distance is Hubble's Law.

Building on work by Vesto M. Slipher done between 1912 and 1925 at the Lowell Observatory, Hubble, in collaboration with Milton L. Humason, vastly expanded the size of the universe during the 1930s by applying Slipher's findings on the Doppler Effect to light. Hubble's use of the redshift in the spectrum colors of light from distant bodies made predicted distances of several billion light-years possible. Hubble's techniques opened new possibilities for interpreting the physical composition of galaxies, their relative ages, and possible distances from Earth, as well as their rates of movement.

206

Velocity equals age

When Vesto Slipher applied the Doppler theory to galaxies, he discovered that they all showed redshifts; there were no blueshifted galaxies, which meant that the galaxies are moving away in all directions. The universe is expanding. This is the origin of the Big Bang theory. Hubble and Humason enlarged Slipher's study until Hubble was able to conclude that objects of the highest velocity are actually the farthest away. Thus, the redshift technique resulting from the Doppler Effect led to Hubble's law: *that the distance of an object in the universe directly relates to its velocity.* The higher the radial velocity, the greater is the distance. A diagram of Hubble's theory is in **19.8**.

Velocity relates to the Doppler redshift of the object's spectrum, but how does it relate? Cosmologists have sought to discover a constant to apply to Hubble's Law because it would hold the key to the age of the universe. By dividing the velocity of an object, as obtained from its redshift, by the Hubble constant, distance results convert to light-years. On the idea that the most-distant object detected in the universe cannot be older than the light it emits, we deduce the age of the universe.

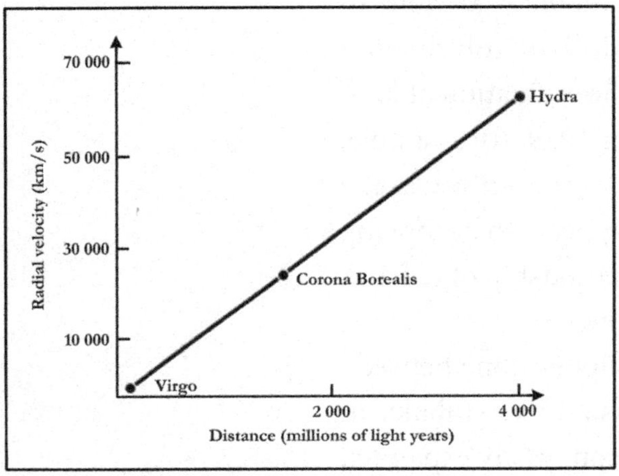

19.8. Diagram of Hubble's Law. Hubble's Law asserts that objects in the universe are receding from us at velocities proportional to their distances.

Determining the Hubble constant is not easy, because there are no reliable distance measurements to far-off objects to use as a yardstick. Estimates exist for a constant of 50 to as high as 100. A constant of 50 implies that the universe is approximately 20 billion years old; a larger constant would make it around eight billion years old. The best value so far is probably somewhere in the middle, making the universe around 12 to 14 billion years old. Hubble initially estimated the constant at 500, but that changed considerably as astronomers confirmed new distances.

For his work, the scientific community honored Hubble when the first major orbiting telescope launched

April 24, 1990, by naming it the Hubble Space Telescope. After a somewhat shaky beginning because of a faulty imaging system, the Hubble Telescope operated superbly after in-space repair, going on to confirm many of Edwin Hubble's predictions about the universe. Operating from outside the atmosphere of Earth, the Hubble Space Telescope expanded knowledge of the universe beyond all expectations. The Hubble Space Telescope in time will refine the value of the Hubble constant, as will the new Webb Space Telescope.

Twelve years after the Hubble's launch, astronauts visited it in 2002 to install a new imaging system capable of a 20-fold improvement over the old system it replaced. The new Hubble has a new resolution so sensitive it compares with an observer in Washington, D.C., being able to distinguish two fireflies sitting side by side in Tokyo. Future discoveries from Hubble the highly sensitive Webb promise to be even more astonishing.

207

Hubble's discovery that the radial velocity of a galaxy away from Earth directly relates to its distance, provided a scale perfectly suited to the employment of spectroscopic analysis. Reading the spectral redshift of light from a distant object makes it possible to estimate the radial speed, or velocity, at which the object is moving away. Once we know velocity, the Hubble graph gives predicted distance.

208

Spectroscopic analysis is the lone technique for estimating distance to very distant objects, objects that are so far away that the only criterion of reliability is that sufficient radiation be detectable to present a spectrum of the body's elemental, chemical composition. All that analysts need to measure the most-distant objects in the universe is a detectable spectrographic code. Moreover, because in astronomy we measure time in distance, the distance light travels in one year, the distance of an object gives its age. Astrophysicists assume that the

19.9. At the edge of the universe. R. Williams, Hubble Space Telescope Deep Field, NASA.

most-distant object is the oldest; therefore, an object predicted to be 14 billion light-years away tells us that the universe must be at least 14 billion years old. In Figure **19.9**, a distant galaxy (red dot at the center of the image) may lie at the edge of the known universe. It is one of the most distant objects ever observed. Larger objects in the picture are at closer distances.

209

Stepping-stones

Predictions of distances to galaxies make it possible to infer even greater distances by comparing similarities among galaxies. Just as in the early beginning of stellar distance estimations, we presumed that all stars were like the Sun; other galaxies we likewise presume to be like Andromeda or other known galaxy types at known distances. In this way, it is possible to jump from galaxy to galaxy, covering in the process unimaginable voids of space and distance. Obviously, such practice carries a large amount of uncertainty. Chéseaux who first put a number to the speculation about stellar distance made an 85 percent error in calculating the distance to nearby stars by assuming that Mars was like a star. Chéseaux had no way of actually proving that Mars was not a perfect reflecting body, although he should have known better. Likewise, modern astronomers have no way of proving that stars are truly alike or different. The same goes for galaxies. Nevertheless, modern astronomy expects to do much better than Chéseaux at estimating distance because of a relentless insistence on proof in an era when proof is much more likely to be forthcoming.

These then are some of the types of distance estimation techniques: parallax; the magnitude of stars and the application of the inverse square law; spectroscopic parallax, or the determination of star spectral types relative to their stellar magnitudes and temperatures; and statistical parallax, whereby distances are calculated from stellar and galactic motions.

210

Each method of distance estimation has its supporters and detractors. Not all stars and cosmic objects turn out to be what they appear to be. Nothing fits conveniently into predetermined niches. There are main sequence stars, different kinds of variable stars, many types of galaxies, and objects that produce radiation that confuse traditional spectral analysis. Take into consideration, also, star and galaxy alignments, proximity of objects that confound both magnitude and spectral analysis; add to this the proper, or sideways, velocity of objects; and one can see that there is good reason to be mildly skeptical about any method that claims to predict the alleged size of the universe.

Most estimates of stellar distance carry a degree of uncertainty. However, when assessing the vast distances of space, errors have a comparable degree of irrelevancy. Even if the degree of uncertainty of measurement is as high as 50 percent, the universe still is enormous. Fifty percent of the average estimate of its scale of 14 billion light-years is seven billion light-years. The uncertainty error is probably realistically less than 50 percent, because Earth itself is more than four billion years old, and the universe is much, much older than the solar system. Unless, of course, we ultimately discover that the cosmic principle does not apply beyond certain distances; that is, that time and space are not the same everywhere. Then all bets are off.

Total unreliability of the cosmic principle is unlikely to be the case either, at least not in a way that would drastically revolutionize distance prediction technology. Estimates of the scale of the universe appear to be credible within reasonable bounds, wanting only of periodic refinement by new instruments and new discoveries, confirming either present understanding or modifying the calibrations of traditional models of measurement.

211

Is an object what it appears to be? Does it behave the way astronomers expect it to behave? Much of the work of astronomy aims at resolving these questions. These inquiries, of course, are at the center of controversy and excitement about pulsars, quasars, and black holes, cosmic bodies that are like nothing else known before. As science understands more about these objects, and as they become more totally integrated into the fabric of the known universe, they are sure to provide an additional calibration for the growing number of techniques for predicting the size of the universe.

Archimedes said, "The 'universe' is the name given by most astronomers to the sphere the centre of which is the centre of Earth, while its radius of which is equal to the straight line between the centre of the Sun and the centre of Earth. This is the common account as you have heard from astronomers." (Archimedes, Sand-Reckoner, 3rd century B.C.E.)

Barely 2,000 years have passed since Archimedes wrote those words that placed the Sun at the edge of the universe. Less than one hundred years ago, the universe existed interpreted to be synonymous with the Milky Way Galaxy. Today, we know the universe stretches beyond the Milky Way Galaxy for billions of light-years, comprised of millions if not billions of galaxies like the Milky Way. In another 2,000 years today's estimates of the size of the universe may prove to be equally shortsighted.

Pierre Brebiette. *Two Philosophers Watching an Eclipse*. 1615–42. Etching, 4 7/16 x 6 5/16 in. (11.3 x 16.1 cm.) The Elisha Whittelsey Collection, The Elisha Whittelsey Fund, 1975. The Metropolitan Museum of Art.

Five Galaxies of Stephan's Quintet. Colliding galaxies pull and stretch each other in a gravitational dance. Webb Telescope. NASA, ESA, CSA, and STScI.

XXI

UNCERTAIN PROBABILITIES

212

The conquest of the vast distances of the universe is by no means final. In the 20th century, millions of light-years — trillions of miles — expanded what once we thought to be its outer limits. At the end of the 20th century the rough estimate of the size of the universe, while imprecise, was arguably more refined in its accuracy than at any previous time in history. Nevertheless, it is almost certain that an estimate of an even greater scale than that now understood is on the horizon, as newer, more powerful tools search beyond the present limitations of known technology.

Yet the prospect of precise measurements of distances beyond the vicinity of our own galaxy is problematic. This conclusion stems from the assumptions underlying prediction of cosmological distances and the uncertainty of those assumptions proportional to the distances in question. That is, the greater the distance, the less reliable is distance prediction.

The methods of assessing distance in the universe are subject to the many yet unknown features of an ultimately vast and highly complex space domain. The techniques of cosmic measurement, ingenious as they are, are only as reliable as the assumed behaviors of the objects measured; that is, an object's behavior must be similar to known experience in order to be 100 percent predictable. Therefore, we may never know the true extent to which any of the methods of determining distance are totally reliable and valid.

The estimation of cosmic distance rests on the simple premise that objects of similar properties will be the same wherever in the universe they reside. As we know, this cosmic principle may or may not hold in all cases. Already astronomers have discovered, for example, that distant galaxies appear different from those that are closer to us. Such a discovery seems logical when one recognizes that our experience with closer galaxies rests on much younger galaxies than those at the outer edge of the universe. We do not have old galaxies in our neighborhood for comparison. Much of the technique of predicting distances to far away galaxies is dependent on calibrations relative to known behavior of younger galaxies. It is not surprising, then, that more distant galaxies appear to behave differently from their closer counterparts. Presumptions about distant galaxies based on similarities to nearby galaxies may not work. It should be pointed out, at the risk of some confusion, that the terms "old" and "new" often used to describe galactic relationships are relative terms. Galaxies at the edge of the universe, while old in their own time-space intervals appear as new on Earth, newly formed bodies as they appeared shortly after the Big Bang. The light just now reaching Earth began its journey when these galaxies were in their earliest beginnings.

213

One of the great questions about the condition of the universe is whether gravity, a necessary construct of the universe, is sufficient to arrest its outward expansion. The universe is expanding from the Big Bang. Physicists calculate that it will take 100 times more matter than is believed to exist in the stars and galaxies for the universe to reverse its expansion and begin its collapse back toward its center.

The density of the universe, that is, the distribution of matter in space, is extremely small when compared to terrestrial, stellar, or even galactic densities. Density is determined by dividing the available mass by the known volume of space occupied. The problem in calculating the density of the universe is that no one knows what the volume of the universe is. If it turns out to be infinite, then the limit of its density is zero. Under such conditions, either it has no density or its density is negligible. In either case, there is insufficient mass, i.e., gravity, to close the Big Bang, and the universe will continue to expand. Astronomers currently accept a density of the universe as about 1E-31 grams per cubic centimeter. Put another way, one gram of mass to be of comparable density would require a volume equal to a million billion cubic kilometers. Imagine a grain of sand inside a cube that is 200,000 kilometers on each side and you have some idea of the great voids of the universe. Ninety-five to 98 percent of the universe appears to be empty.

214

Voids

New stars form constantly wherever conditions are right. In addition, stars routinely die, to become other different kinds of stars, such as white dwarfs or neutron stars. Alternatively, they may explode and disintegrate entirely into cosmic recycling.

Most of the mass of the known universe is in the stars. Yet 90 percent of the mass known to exist is invisible; it produces no light.

Enormous dark halos of radiation surround visible objects. Often, only a fraction of radiated energy is within the visible range. Many objects are detectable only by the energy they radiate outside the visible spectrum.

The darkness of the universe takes many forms. Boötes Void, for instance, is a 330 million light year diameter empty spot of the universe where only a handful of galaxies appear to exist. Boötes Void is often confused with Barnard 68, a large area of the universe that also looks like a void but is instead a dark molecular cloud that shields parts of the universe behind a high

concentration of dust and molecular gas that absorbs practically all visible light emitted from background stars. Barnard 68 and Boötes Void are two different void phenomena (**20.1**).

20.1. Barnard 68 molecular cloud. Barnard 68 is a dark absorption nebula towards the southern constellation Ophiuchus, within the Milky Way Galaxy at a distance of about 400 light-years. European Southern Observatory.

Before space-based telescopes, much of the light reaching Earth from outer space escaped detection, deflected or absorbed by Earth's atmosphere. Add to this now the knowledge that substantial quantities of light are being gobbled up by massive black holes, and one easily can conclude that the visual record of the universe is an infinitesimal speck of its reality. At the same time, however, where the universe appears to be empty, it may not be empty at all.

215

The light in the universe appears to be collected at specific points. Stars occupy no uniform pattern from galaxy to galaxy, or within a galaxy. Their occurrence is altogether random. Galaxies, on the other hand, appear remarkably regular in their shapes. Moreover, the location of the galaxies gives the universe itself a uniform pattern, at least what is visible from our vantage point in the Milky Way. Clusters of galaxies appear to congregate at points separated by enormous distances. The universe looks like a sponge, clusters of galaxies separated by great voids. The photograph in **20.2** of the northern two-thirds of the sky shows all galaxies in that quadrant of the sky that are brighter than the +19 magnitude. Galaxy clusters and filaments of galaxies joining the clusters appear in the sky, along with the great voids between them.

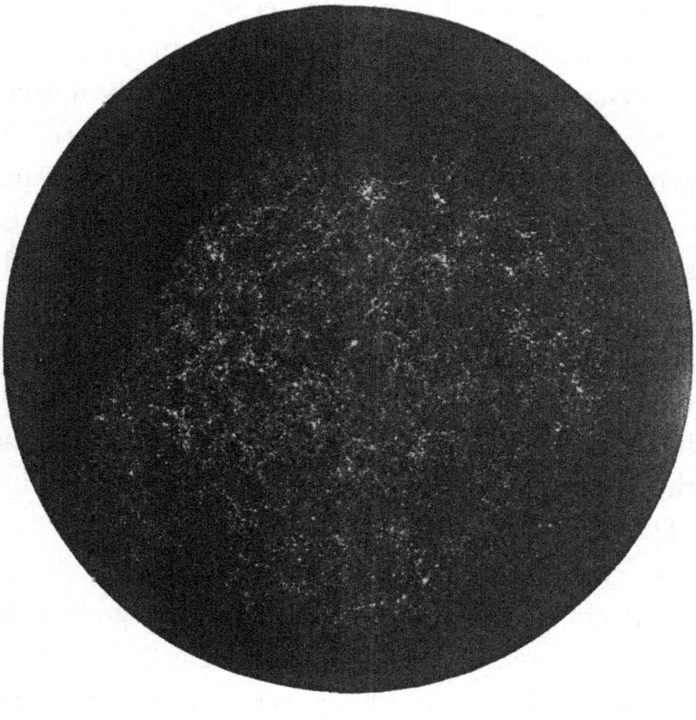

20.2. The voids of space. A photograph of a large section of the universe reveals a web of galaxies interspersed with voids. Lick Observatory.

Still, such large-scale views of the universe have not yet resolved the question of the assumed homogeneity and isotropy of the universe. An isotropic universe would be identical in all directions, which one might assume to be the case as the result of the Big Bang, but so far, the universe is not the same in all directions. On all scales where adequate optical data are available, the distribution of matter appears to be inhomogeneous and anisotropic. The only way it can be isotropic is if homogeneity comprises a construction of both visible and invisible matter.

216

Dark matter — a kind of matter different from black holes and the remnants of dead star systems — was first a topic of theorists in the 1930s but ignored by the astronomy community until in the 1970s. Debate continues on whether or not it exists. Astrophysicists cite anomalies in gravitational fields as proof of the existence of dark matter. Such *ad valorem* processes, however, are not widely accepted as incontrovertible proof.

Nevertheless, the question of dark matter is critical to understanding the universe. If it exists then the universe is possibly closed; if not then, it is most likely open. The primary focus of the search for dark matter is, of course, on the apparent emptiness of space, where no light is seen and where if additional matter is to be discovered it must be in a form unknown and so far undetected. What, if anything, exists unseen in the great voids of space? Methods of reading the universe, other than the interpretation of light, exist, such as radio telescopes, but these, too, are subject to many of the same shortcomings of visual detection.

217

Relativity

Albert Einstein, if he taught us anything about the universe, taught us that things are not as they appear to be necessarily. In his 1905 paper on special relativity, Einstein drew us toward new ways of thinking about the universe and away from reliance on Newtonian space and time. Instead of immutable frames of reference, such as Isaac Newton envisioned, Einstein showed that time and space are relative.

The concept of relativity rests on two things: (1) the speed of light is the same wherever we observe it, and (2) physical laws apply within a constant frame of reference. The question underlying Einstein's theory of relativity is what happens when the frame of reference is not constant for all observers. He realized that the only way these two postulates — universal speed of light and constant frame of reference — could be true was if the time-space intervals changed proportionally to the speed of the frame of reference relative to the observer. In other words, the velocity of the frame of reference determines the time-space interval. At high velocity, the time-space interval in a frame of reference must increase relative to the speed of light observed and equal to a frame of reference at rest. A clock ticks more slowly traveling at high speed than does its counterpart sitting stationary. Astronauts who travel to the Moon and back return younger than they would have been had they remained in Earth's frame of reference. In 1916, Einstein expanded his special theory of relativity to include gravity. When *generalized* to include gravitation, the equations of relativity show that the shape of space-time when under gravitational force not only shrinks or stretches in interval, but matter will cause the flow of time to slow down and change spatial directions. A black hole, for example, if visible would appear to an outside observer to exist in arrested time. Within the black hole's frame of reference, however, time and space would appear normal. From the outside, the observed time-space interval at the event horizon of the black hole is so large that nothing appears to be moving.

The universe is a minefield of massive bodies, each with the capability to warp time and space into a convoluted space scape of zigzagging light traps with the potential to distort radiant energy and its record of the emitting sources. It is hard to know if the energy received in Earth's frame of reference is the same uninterrupted energy that left the relative frame of some distant galaxy.

218

The speed of light

Light travels at approximately 300,000 kilometers per second in a vacuum. According to Einstein's special theory of relativity, this is the upper limit to the velocity of the propagation of luminous energy. Nothing can travel faster. (Laboratory claims of a 7 percent increase in the speed of light in the 1980s, using so-called super-luminous propagation, turned out to be a circumstance of wave behavior and not an actual increase in light speed.)

Just because there is an upper limit to the speed of light, it does not necessarily infer a lower limit. Light can travel slower than 300,000 kilometers per second, much slower. In fact, one of the remarkable qualities of light is its ability to slow down when entering a denser medium, only to return to its original speed upon exiting the medium.

Most of the experiences with slow light come in the form of *refraction*, or the bending of light through a glass prism or other transparent medium. Glass being denser than air, light slows and deflects relative to the different frequencies contained in the light, thus causing it to exit the prism in the familiar spectrum of different colors.

In 1999, an investigative team of scientists slowed light to 1 kilometer per second, thus reducing the speed of light by a factor of 300,000. This remarkable feat occurred by directing a beam of light through an artificially induced electromagnetic transparency made up of overlapping atoms at ultra-low temperatures. This new type of matter is Einstein-Bose condensate, a material theorized in 1924 by Einstein and India physicist Satyendra Nath Bose.

The universe is both a vacuum and not a vacuum; a vacuum in the sense that space contains no air, but at the same time not a vacuum because matter is all over the universe. Most likely some of this matter is in the form of Einstein-Bose condensate. The new light-slowing atomic matter produced in the laboratory was super cooled to within 50 billionths of a degree above absolute zero, absolute zero being the temperature at which atoms are completely frozen and cease to interact in any kind of energy transformation. The nearly absolute temperature reached in the laboratory is the lowest temperature ever achieved and cooler, physicists believe, than anything in deep space. Yet, who knows what path light may take on its long journey through the universe to Earth? If light is subject to alteration experimentally, the possibility exists that it will be altered naturally. Studies by the same group of scientists who reported the dramatic results above, slowed light in an earlier experiment to 3200 kilometers per second, about 1 percent of its normal speed, just by shining it though a cloud of illuminated lead atoms.

New matter environments may be common in unexplored regions of the universe. Light passing through such environments could bring a very different message about the age and size of the universe when traveling at a velocity below 300,000 kilometers per second, especially if the velocity is anything approaching those slower speeds already demonstrated on Earth.

219

Gravitational lenses

Within Earth's solar system, the general theory of relativity as observed holds true relative to the planets. We learned years ago that reflected radiation from a planet, when the radiation passes close to the Sun, bends to the extent that a planet can appear to be where it is not. On a cosmic scale, this distortion factor greatly magnifies. When examining very distant objects, the use of an earth-based frame of reference to interpret observations of extra-galactic space risks all manner of error by ignoring the general theory of relativity.

One small example of this is gravitational lensing, a space warping phenomenon that not only causes distant

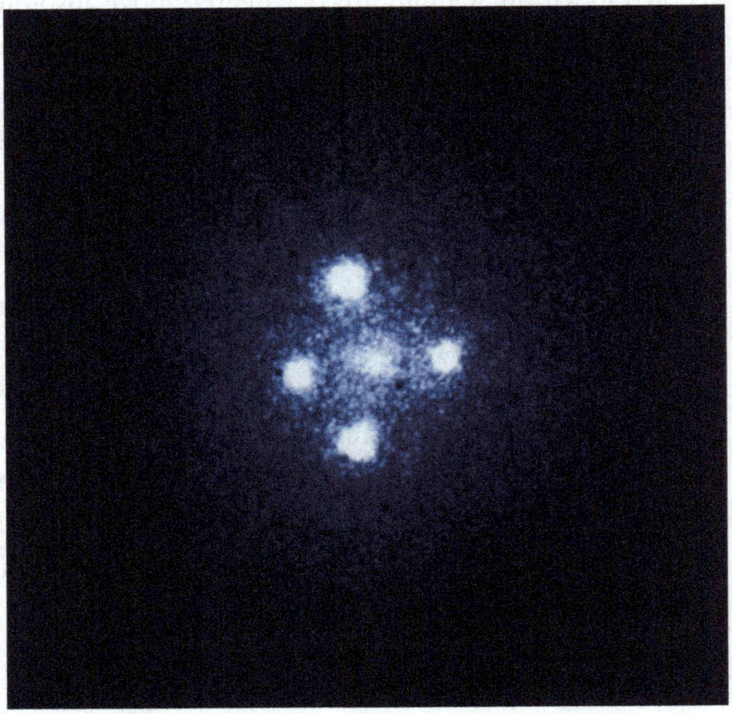

20.3. The Einstein Cross. Gravitational lensing causes four images of the same distant quasar whose light is bent by a nearby galaxy acting as a gravitational lens. NASA, ESA, STScI.

objects to appear to be where they are not, but also magnifies them at the same time. The Hubble Space Telescope accidentally discovered in the process of looking for other cosmic phenomenon, a collection of patterns — rings, arcs, and crosses — that turned out to be projected images made by gravitational lenses. A gravitational lens acts like a giant magnifying glass that enables objects to be visible that are otherwise beyond detection.

The first lens system and the brightest came to attention in 1995. Ground-based telescopes confirmed the lensing galaxy shown in **20.3** seven billion light-years away. Its gravity magnifies the quasar behind it in multiple images making the quasar visible although it is about 11 billion light-years distant. The probability that similar time-space distortions exist in many forms throughout the universe adds to the risk of drawing absolute conclusions about the uncertain nature of the cosmos. Much about the universe remains unproven. For the time being, we must accept that we are inside a dynamic, changing space about which we know very little.

Gerrit Dou. *Astronomer by Candlelight.* late 1650s.
Oil on panel, 12 ⁵⁄₈× 8 ³⁄₈ in. (32 × 21.2 cm.)
The J. Paul Getty Museum, Los Angeles.

Anonymous, after Parmigianino. *Astrology*. 17th century. Etching, 6 5/16 x 4 1/8 in. (16 x 10.5 cm.) Harris Brisbane Dick Fund, 1927. Metropolitan Museum of Art.

XXII

LIFE IN THE UNIVERSE

220

Those who inquire into the nature of the universe turn inevitably to the question of life on other planets, seeking the possibility of life not limited to the isolated confines of Earth.

Among the billions and billions of stars in the universe, could conditions somewhere else support life? Simple laws of probability would say yes, many times over. The average life span of a star could include one or more attracted masses of matter of sufficient size to be a planet, held in a uniform orbiting pattern within the star's gravitational field for a sufficient length of time to develop the ideal conditions necessary for life. However, we know only that in the vastness of the universe the Sun alone possess a planetary system like ours. Nevertheless, it is unlikely that Earth is unique in the entire universe.

For most of the history of astronomy, the ancient astronomers thought Earth was at the center of existence; later, the Sun became the center; and the galaxy at one time was synonymous with the universe. Time and new discoveries erased these early misconceptions. It is unreasonable now to believe that Earth is the only living planet in the universe.

221

Archaean life

At the beginning of the 19th century, all living organisms conveniently organized in two kingdoms: animal and plant. In time, it became clear that life was more diverse than just two classifications could adequately express, and various other kingdoms arose that proposed to divide life into five classes. Microbiologists grouped these five kingdoms under three basic varieties of life: archaea, bacteria, and a single classification of higher life forms devoted to plants and animals.

A good deal of knowledge exists about bacteria, as well as the many species of plants and animals. We know much less about archaea and the Precambrian era from which these life forms emerged.

21.1. Morning Glory Pool during winter. Yellowstone National Park.

Ninety percent of the development of life on Earth occurred during the Precambrian age, the oldest and largest division of geologic time. Precambrian Archaean rock is the name of the oldest rocks known on Earth. It is in these rocks that geologists discovered the fossils of Archaean life. They are the oldest known fossils of life and very similar to microbes still thriving today. Modern Archaean life forms have been discovered living in the most extreme and uninhabitable regions of Earth. Archaea grows abundantly, for example, around the hot thermal springs of Yellowstone National Park (**21.1**).

Bacteria and archaea have amazing qualities. Each probably developed from a common ancestor four billion years ago. They can literally go into hibernation for millions of years, remaining in suspension under extreme conditions. Microbiologists once revived a bacterium from the stomach of a bee trapped for 25 million years in a chunk of amber. When freed from its spore-like state, the bacteria revived quickly and grew.

When bacteria were discovered living under conditions so inhospitable as to question their ability to survive, scientists began looking more closely at their characteristics, thus was born a new domain of life, similar to, but different from, bacteria: the archaea. Archaea are biochemically and genetically different from all other life forms. Their habitats include the extreme environments of the planet. They live near volcanic sea vents at temperatures above 100 degrees Celsius (212 degrees Fahrenheit). Others are in an icebound Antarctic lake 4 kilometers (2.5 miles) below the ice surface. This lake, incidentally, may contain other life forms that are millions of years old.

21.2. Hubble Space Telescope photo of Mars taken when the planet was 50 million miles from earth. Bell and Wolff, NASA, ESA.

There is fossil evidence of an Archaean microbial colony on the primal Earth more than 3.5 billion years ago. Earth at that time would have been a hot, inhospitable place.

The survivability of these simple life forms fuels new speculation about the possibility of microbial communities on other planets.

In 1996, NASA stunned the world when the agency announced the finding of bacteria-like fossils, similar to archaea discovered in Precambrian rock, in a martian meteorite. The meteorite, recovered from the Antarctic continent, is a significant find in modern astronomy if its evidence of life forms turns out to be correct. So far, though, conclusive proof of life on Mars awaits closer examination of the red planet sometime in the future (**21.2**). If life ever existed on Mars, it left a fossil record. Meanwhile, the similarity of the ages of the Martian meteorite and Earth fossil record of 3.5 billion years encourages imaginative observers to muse about the *seeding of life* on Earth from extraterrestrial origins.

Archaean microbes can coexist with precipitating minerals and remain preserved in that state for hundreds of millions of years. They literally grow with the formation of the rock itself. Archaea thrive in extremely alkaline or acid waters, or in very salty water. They eat hydrogen gas, carbon dioxide, and sulfur. Some of them use Sunlight to manufacture energy.

Until recently, life elsewhere in the universe was a statistical probability, but one that was difficult to imagine except in science fiction stories because of the known inhospitable environments of other planets in the solar system. Now that we know that many archaea thrive and multiply in conditions that would destroy other living things, hostile extraterrestrial environments become less of a limiting factor to life beyond Earth.

The fundamental characteristics of bacteria and archaea as single cell, non-nuclei life forms are nearly as old as Earth itself. Multicellular life, which accounts for all the other species of life, did not appear until relatively late, some 540 million years ago. Moreover, the first bipedal hominid only emerged a little more than three million years ago. A clear distinction, then, is that of microbial life versus advanced intelligent life. The scientific search for extraterrestrial life within and beyond our solar system primarily involves microbes. Still, other possibilities exist.

222

Hominid planets

The earliest remains found of true humans are from Africa. These remains date in the neighborhood of 2.3

million years old. Hominid tools date from even earlier, and anthropologists believe *Homo habilis*, the first human, originated around 2.5 million years ago during a period of rapid environmental change on Earth.

The natural tendency when looking for extraterrestrial life is to look elsewhere in the universe for conditions that replicate Earth experience. One of the essential requirements, it is assumed, is the existence of a star system like our Sun, with at least one planet like Earth, or perhaps Moons that are Earthlike in their environments.

The existence of life is dependent on, among many things, the temperature of the environment. Not all life requires the same temperature. Life can exist above and below the 98.6 degrees Fahrenheit associated with normal human life. The extremes are limited, however, to a narrow range, and are entirely dependent on the temperature of the star serving the planetary life system and its distance from the life-supporting planet. The surface temperature of the Sun, for example, is about 5,500 degrees Celsius (10,000 degrees Fahrenheit). Its distance at 150 million kilometers creates the precise environmental conditions to sustain life, as we know it. This is the life zone. By comparison, the hottest stars are 50,000-100,000 degrees Celsius (90,000-180,000 degrees Fahrenheit) far too hot to sustain life except perhaps on very distant, slowly orbiting planets in life zones much farther out than the life zone of our Sun. In addition, hot stars burn up comparatively soon and for that reason are not viable life-support systems. Therefore, in the search for intelligent life, the objective is to look for planetary systems similar to our own. There are likely billions of stars in the universe of the same temperature class as our Sun with life-zone potential like ours.

223

The discovery of new planets in other star systems is a late 20th century addition to astronomy. In the 1960s, suspicion that the often-mentioned Barnard's star had a planetary system never materialized. After a hiatus of 30 years, claims surfaced in the early 1990s for new planets around pulsar stars.

Planets beyond the solar system are too small to see. The Webb Space Telescope is capable of examining the atmosphere of large exoplanets. However, we detect planets outside the solar system primarily by measuring their interaction with their host star. Where there is no detectable light or radiation of any kind, the presence of an object reveals itself by its gravitational effect, or by its eclipsing interaction with its host star. Although planets are outside the range of visible confirmation, the slower rotation of some stars is due to the gravitational drag of unseen planetary systems.

The first planetary system discovered in this way was in 1992. This system orbits an invisible neutron star. Because the star emits no light, the planetary system is obviously radically different from our own.

Eventually, in 1995, planetary systems began to appear detected around normal stars. Two such stars are 51 Pegasi, in the constellation Pegasus, and 47 Ursa Major, home of the Big Dipper. The star 47 Ursa Major is visible with the unaided eye in the springtime evening sky.

By the end of 1995, astronomers discovered 20 extra-solar planets, in addition to the discoveries of 51 Pegasi and 47 Ursa Major. Since then, we know that literally hundreds of planetary systems exist. Still, planets in most of these systems are more like Jupiter than Earth, and while they orbit main sequence stars, the stars are rarely like the Sun.

Good fortune rewarded the search for a sun-like system in 1999 when a research team at Lick Observatory discovered three large planets in a system of a sun-sized star in the constellation Andromeda. The star is sun-like, but the planets, like those previously identified, do not appear to be earthlike. They are Jupiter-size and larger, and are orbiting very close to the parent star, well inside the star's life zone.

Astronomers at the Keck Telescope quickly followed the Lick discovery with the detection of six more jumbo planets orbiting stars, again much like our Sun. In this instance, however, five of the planets have temperatures predicted to be earthlike.

As the possibilities of hominid-like, life-sustaining environments mount with the discovery of more and more new planets, and as some astronomers concentrate on developing the technology to locate smaller, earth-size planets, others remain involved in the search for more primal life forms and are concentrated on locations closer to home.

224

Water

Life requires water if it is to propagate. Water is the most abundant substance on the surface of Earth. So defining is its presence it becomes a base reference when calculating density of elements and when measuring temperature. Essential to life on Earth, water covers 70 percent of the planet, and is necessary to life everywhere.

Bacteria and archaea can thrive in almost any kind of liquid environment for millions of years, even under extreme conditions. Hominid intelligent life, on the other hand, is more precarious.

Ninety-five percent of all water on Earth is in the form of salt water, which cannot sustain human life; only 5 percent of Earth's water is drinkable and more than half of that exists locked up in glacial ice. Of the remaining 2.2 percent of fresh water after pollution and other factors come into consideration less than 1 percent of all Earth's water is fit for human consumption, and that is diminishing steadily.

At one time, scientists thought that water might be unique to Earth. However, 20th century discoveries proved otherwise. Water exists on the Moon, lies in abundance beneath the surface of Mars, and is the main ingredient of comets in the form of ice. There is evidence that water once flowed on the planet Mars and photographs suggest that Mars at one time had abundant lakes and perhaps seas. So copious is water in and around the solar system that some think the great amounts of water on Earth may have come actually from outer space, deposited over billions of years by passing comets or other water-laden bodies.

It is no longer farfetched to imagine Archaean-like microbial communities in other places beyond our own planet. At one time water stood out as one of Earth's unique life-producing, life-sustaining characteristics. It is now being revealed that water is probably everywhere in the universe.

The Greek astronomer Thales of Miletus would be pleased. He believed 25 centuries ago that water was the primary element of the universe. That distinction, of course, belongs to hydrogen. Nevertheless, his instincts were on the right track. This same Thales showed the Egyptians how to measure the height of the Great Pyramid by measuring its shadow. Renowned for his wisdom, many consider him the first philosopher in a time before the term philosopher was common.

225

While most extraterrestrial planets because of their size are probably not earthlike — they are more like Jupiter — a planet in the habitable life zone of a star may have moons, just as the planet Jupiter does, where the possibility exists that a benign environment could support the emergence of life. For example, there is evidence of a liquid ocean on the Jupiter moon Europa, just below the moon's frozen crust (**21.3**).

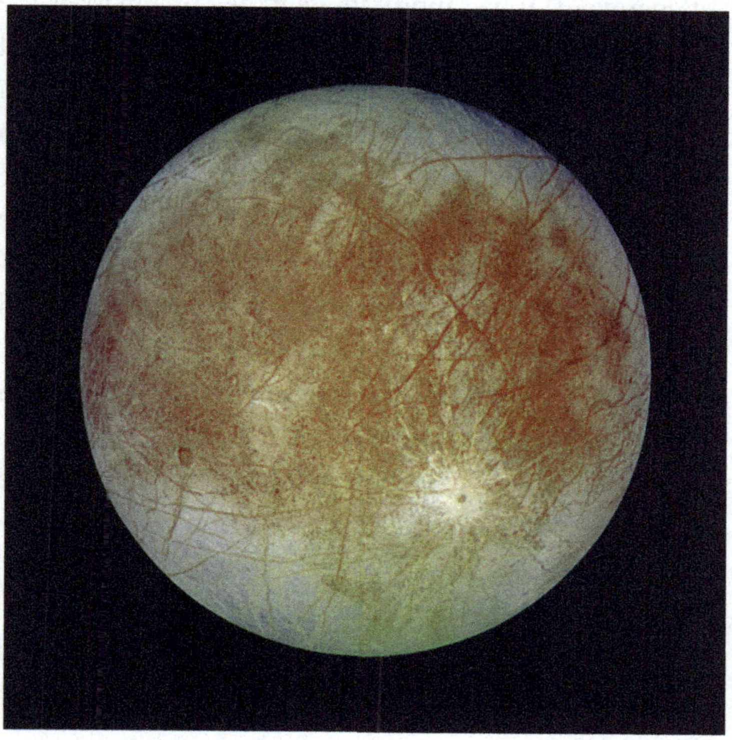

The tidal forces of Jupiter have rendered the surface of Europa in telltale cracks that appear rimmed with a brown discoloration. No one wants to say, but this staining has much of the appearance of archaeal-bacterial deposits. Answers will have to wait until a future space mission visits Europa to take a closer look at its subsurface ocean. The bright spot in the lower right corner in **21.3** is a recent impact crater 50 kilometers in diameter. Europa itself is about the size of Earth's moon.

21.3. Jupiter moon Europa. The Jet Propulsion Labortory, Pasadena, CA; Galileo Mission, NASA.

226

The most likely candidate for finding life like us is in a relatively nearby setting, our closest star neighbor Alpha Centauri. Alpha Centauri passes at least four of the 5-criteria test for terrestrial life, as we know it. (1) Alpha Centauri is the same color and spectral type as the

Sun. This means it is a main sequence star; hence, it is mature and stable like the Sun; (2) it is the same temperature as the Sun, emitting about the same amount of energy; (3) it has approximately the same mass as the Sun, being just slightly larger; therefore, its lifespan and future are predictable; and (4) life in the Alpha Centauri system has had time to evolve. Alpha Centauri is five to six billion years old, old enough for life to develop. The Sun by comparison is about 4.6 billion years old. So, life in the Alpha Centauri system, if it exists and if it developed on the same time scale as on Earth, could actually be more advanced than life on Earth

We do not know if water exists in the Alpha Centauri system, or if planets that may exist in the Centauri life zone have atmospheres. Both water and atmosphere are necessary elements of the fifth criterion for life. Since water is a major source of oxygen, we may assume that if there is sufficient water to sustain life, there may also be some kind of atmosphere to shield life from harmful energy coming from outer space, just as our atmosphere shields Earth.

The illustration in **21.4** diagrams Alpha Centauri A and Alpha Centauri B both of which meet most of the

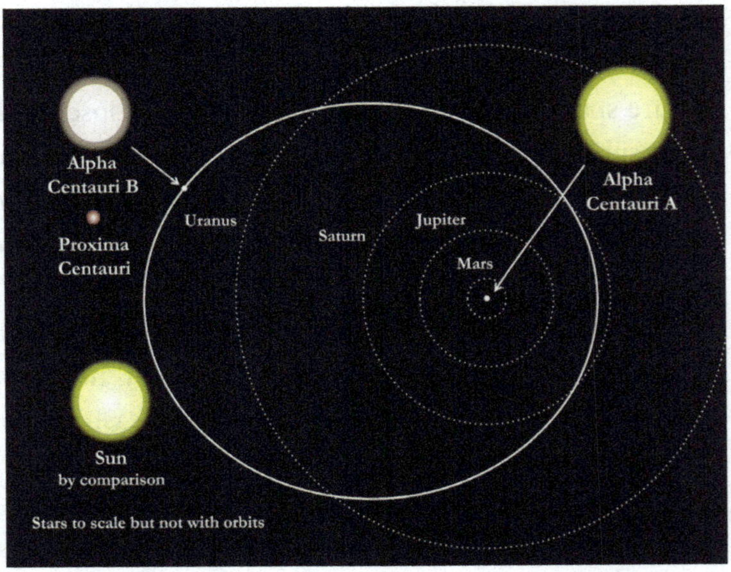

21.4. Centauri star system compared to the Sun and solar system. Planetary Habitability Laboratory, University of Puerto Rico at Arecibo.

criteria for terrestrial life because of their similarity to the Sun. Orbits of some of the planets in our solar system show the relative scale of Centauri A to Earth's life zone inside the planet Mars. Proxima, or Alpha Centauri C, the third star in this triple star system, is to scale in the drawing but lies well beyond the orbits of Alpha Centauri A and B, and does not meet the criteria for life, as we know it.

A life-sustaining Alpha Centauri planetary system presents a uniqueness complicated by the binary nature of the system. Alpha Centauri is in reality a three-star system. Including planets in an already complex star system would be a tricky balancing act at best but is by no means impossible. Therefore, astronomers continue to look at this nearby star system as a likely candidate for intelligent extraterrestrial life.

The search for extraterrestrial life is the most exciting, yet potentially the most sobering, undertaking in modern science. In 1996, a high NASA official predicted discovery of extraterrestrial life in some form in the next 25 years.

227

So far, life on Earth has managed only to sputter along from one mass extinction to another without much to show for it, except perhaps the tools to engineer yet another mass extinction at our own hominid hands. Yet the discovery of new life could happen, providing Earth survives its own encounter with life long enough to search for others.

The question of intelligent life beyond Earth is in some respects mute. Hominid life and the structure of the universe are incompatible constructs. Qualities that define evolved life, that is, physical places, interaction of species, and mortality, have no applications in the time-space intervals of the cosmos. Outside the solar system, the universe becomes an abstraction. Intergalactic travel is impossible within the confines of our physical reality.

Consequently, alien communication and interaction beyond the solar system has no meaning in the ordinary context of a real-time exchange of experiences. At the same time, the incomprehensible distances in the universe render impossible any idea of a common space — a sense of place. In terms of human understanding of the universe, where time is at once present, past, and future, the cosmos is a place of inescapable non-objectivity.

228

Stories of giant icebergs the size of the state of New Jersey, breaking away from the Arctic ice shelf to float freely in the vast Atlantic Ocean, stand us in awe of the power of nature. The mind does not fully grasp events of such magnitude. Yet, in our perception of the solar system — the Sun and the planets — we accept the fact that enormous objects — the orbs of space — move suspended in harmonious patterns guided by natural forces of enormous power. We extrapolate from these facts a crude knowledge of the greater cosmos, but the reality of the greater universe is incomprehensible. If massive, free-floating chunks of ice adrift in the ocean are almost unimaginable, think of a tiny grain of sand inside the space of a human footprint, and that is the approximate relationship of the size of Earth to the size of the Sun. Think then of the same earth-sun footprint made on Earth. The minuscule imprint of one tiny footprint on the immense surface of Earth is in the same proportion to Earth as the imprint of the solar system on our galaxy, and our galaxy is only one of millions of galaxies in the universe. No person can experience the true physicality of the solar system. The human mind cannot grasp the scale of the Sun and Earth, let alone the galaxy and beyond.

How then are we to know the universe? Cosmic experiences are indescribable by human perception. Therefore, our imagined view of the universe entertains the soul. We exercise our being when we try to figure out what and how the universe is.

Navajo Tséyi, Canyon de Chelly National Monument in Arizona. NASA/JPL-Caltech.

Mars' Jezero Crater Delta, in natural color. NASA/JPL-Caltech/ASU/MSSS.

William Blake. *The Great Red Dragon and the Woman Clothed with the Sun.* ca. 1805.
Pen and gray ink with watercolor over graphite, 16 1/16 x 13 ¼ in. (40.8 x 33.7 cm.)
Rosenwald Collection. National Gallery of Art.

THE LOGOS AND THE COSMOLOGICAL DIALECTIC

And whether it is clear to you or not, no doubt the universe is unfolding as it should.

Max Ehrman.
Desiderata, 1927.

William Morris Studio. Acanthus Portière. 1890s, Mary Isobel Barr Smith, embroiderer.
Silk embroidery on linen, 9 ¾ x 6 ½ in. (24.7 x 16.7 cm.)
Art Gallery of South Australia.

XXIII

THE IDEA OF A UNIVERSE

229

Throughout many centuries, the greatest minds of history have tried to solve the riddle of the universe. Numerous theories exist for its origin and structure. Yet the great question remains: is the universe owed to natural or supernatural forces?

Until about the sixth century B.C.E., it is almost certain that the theory of supernatural forces was immanent in the world's earliest civilizations. The view that the world represented a natural as well as a supernatural sequence of events appears first to have emerged in ancient Greece, to be developed by Pythagoras and his followers. The concept of an ordered cosmos mathematically described has occupied us ever since as a large part of the human quest to understand the universe. Nevertheless, the finite, geocentric cosmologies of Plato, Aristotle, and Ptolemy balanced against the meticulous, mathematical arguments of Copernicus, Brahe, Kepler, Galileo and Newton never totally displaced the ancient and medieval theologies of an infinite realm created by the hand of God.

230

Truth and belief

Some of the most profound epiphanies about the universe are in the poetry and philosophies of people from all ages. The quest for truth about the universe is the common motivation. Divinely inspired or agnostic, humans seek to uncover the ultimate context of human existence. It drives us to apply our unique mental capacities, without perhaps knowing why, to discover the purpose of being, and to seek it in places beyond the stars. H. G. Wells said, "There is, though I do not know how there is or why there is, a sense of infinite peace and protection in the glittering hosts of heaven. There it must be, I think, in the vast and eternal laws of matter, and not in the daily cares and sins and troubles of men, that whatever is more than animal within us must find its solace and its hope." Humankind dreams of leaving Earth's plane to travel the universe. Many, including Wells, wrote imaginative tales of doing so. Whether the trip is to be real or imagined, physical or spiritual, determines the course of individual human action.

When in 1969 humankind landed on the Moon, part of one dream was realized, the dream to travel in outer space beyond Earth. Nevertheless, what might then have been a celebrated modern triumph of the natural conquest of space became instead a metaphysical experience. Looking at the beautiful but fragile sphere of Earth from beyond its terrestrial plane inspired both scientists and poets to reflect on this new and previously unknown perspective of life. It was as if going to the Moon was almost like something outside nature's plan, or certainly in addition to it.

American astronaut Gerald P. Carr, who flew Skylab Mission IV, remembered his most vivid impression of being in space and looking back at Earth and the fragility of its thin blue layer of atmosphere, and coming to the sober realization that there was, in his words, "nothing beyond to replace it; only the black of deep space." Archibald MacLeish, the late 20th century poet, penned a similar reaction to the first pictures of the solitary Earth from outer space. He wrote, "To see Earth as it truly is, small and blue and beautiful in that eternal silence where it floats is to see ourselves as riders on Earth together, brothers on that bright loveliness in the eternal cold." MacLeish was responding to a live television telecast from Apollo 8. Apollo 8 was the first manned space flight

to orbit the Moon. As the ship flew over the cratered, gray surface of the Moon on Christmas Eve, the crew took turns reading, heard across the great void of space, the Old Testament Book of Genesis: "In the beginning, God created the heaven and Earth. And Earth was without form, and void; and darkness was upon the face of the deep."

22.1. Planet Earth. H. Schmitt, R. Evans, Apollo 17 crew, NASA.

The photograph in **22.1** is the whole Earth image taken from Apollo 17, the last Apollo mission to land on the Moon, and the first flight to get a full view of our planet. This picture became an icon of the six lunar landing missions. Taken in December 1972, four years almost to the day after Apollo 8 sent back those first

Christmas Eve pictures of Earth, the 1972 photograph showed a cloud-encircled Earth as a blue sphere against the backdrop of space. The view is from the Mediterranean Sea to Antarctica. The continent of Africa, origin of the human species, is set in rust-colored relief against the oceans that surround it. The African coastline is clearly visible, as are the Arabian Peninsula and the island of Madagascar. The Red Sea stands out in stark contrast to the surrounding landmasses. Earth is like a large marble floating in space, with its warm brown Earth tones, dazzling deep blues juxtaposed with the brilliant white of the south pole, all glowing through a lacy filigree of clouds. So stunningly beautiful is Earth from outer space that one of the Apollo astronauts is said to have looked up from the Moon's surface to see Earth shining above, and wept. The distinguished British astronomer Sir Fred Hoyle (who came up with the name Big Bang to describe the expanding universe) had predicted decades before Apollo that once a photograph of Earth taken from the outside showed the sheer isolation of the planet, a new idea as powerful as any in history would be let loose. He was right. It is one of the most compelling pictures ever taken.

The ancient writers of Biblical scripture had it right: In the beginning, the universe was without form. On that, there is full agreement. In the beginning before the Big Bang, there was no space, no time, and no motion. Augustine, arguably the greatest theologian of the Christian faith, summed it up best when he said that there was no "before" and there was no "outside." Time and space started with the creation of the universe.

231

Zeno and Parmenides

The Greek philosopher Zeno of Elea, in his 5th century B.C.E. philosophical examination of the nature of the universe, laid down issues still without resolution in the 21st century. A whole, he wrote, must be composed of ultimate indivisible units, or it must be divisible *ad infinitum* because an infinite number of parts cannot add up to a finite whole. Zeno's dialectic method, as Aristotle called it, sought to arrive at the truth by disclosing contradictions. Zeno began his argument by discussing reality as a chain of multiple events that are, in effect, in motion. Then he showed that motion and multiplicity are logically impossible. A moving body can never come to the end of the line; it must first come half a line, than half the remainder, and so on *ad infinitum*. Common sense notions about time and space are contradictory, he said. This makes sense, especially when applied to the universe. Distant galaxies that appear to be traveling at ever increasing velocities toward an unknown destination have no place to be going, and would never get there if they did. Zeno argued that all change and motion are illusions of the senses, and he denied physical change as a separate reality. Either things exist or they do not. There are no in-between stages, as the concept of change, or becoming, ordinarily implies. The universe, according to science, is changing and becoming all the time, but there is nothing for it to become. The great void into which common sense says the universe is expanding does not exist; its only presence is an illusory one relative to the edge of the universe at a particular point in time, an edge defined not by the truth but by belief.

Parmenides, Zeno's teacher, who is also the originator of the concept of omnipresent being as the substance of the universe, first argued the dualism of truth and belief. Parmenides — himself a member of Pythagorean society — held that it is impossible to recognize not being. He believed that merely thinking about not being was an act of being; therefore, there is no such thing as not being. The Parmenides principle that thought and being is the same thing survived in the often-cited dictum of the 17th century French mathematician and philosopher René Descartes who said, "I think, therefore, I am."

If there is nothing beyond the universe and no space in which it can exist is there then a universe at all? In this way, logic denies any place for the universe to exist as a physical reality, because it cannot be and not-be at the same time. Echoed by Zeno, and it must be said the Algonquin Indians and countless other great religious and theosophical interpreters of spirituality, the ultimate Parmenides reality is one of undifferentiated *being* in contrast to the illusory testimony of the senses.

We find this idea expressed particularly well in the Buddhist *samsara*, the eternal, undifferentiated stream of being. The metaphysics of Buddhism stresses the essential solidarity and oneness of the universe. The individual is not a separate entity but rather a grouping of elements. Similarly, the ancient people of India used the word *prana* to identify the timeless, universal life force. In the great Chinese Han Dynasty, *qi* described the same circulation of vital life energy. The same idea has its counterparts in numerous other historic and modern cultures, including the popular westernized practices of spiritual ergonomics, T'ai chi ch'üan (T'ai Chi) and Feng Shui.

Parmenides was steadfast in his argument that being is the material substance that comprises the universe; it is the sole and eternal reality. To this point, Zeno reasoned that because being extends spatially and is all that exists, there is no empty space in the universe, and motion is therefore impossible. In modern context, what is observed as an expanding universe is nothing more than an illusion that reduces what is multiple in our experience of cosmic phenomenon — stars, galaxies, black holes — to the unity of a homogeneous, systematically organized process. When this process broadens to accommodate the logic of Parmenides and Zeno, the timeless theologies of the world's great civilizations, and the mind-bending theories of Einstein, it is easy to conclude that the universe is not a "thing"; it is an idea; not a place, but more Belief than Truth, an idea within an Idea.

232

The shape of infinity

Some theorists think that Einsteinian cosmology is pervasive in shaping the universe. The explosive outward velocities of distant galaxies, assumed to be due to the Big Bang beginning of the universe, may actually be a manifestation of the general theory of relativity. The galaxies may not be moving at all; instead, they simply may appear to be moving because of the time-space expansion of the curvature of the universe.

Shape is critical to the accurate prediction of the age and size of the universe. One needs only to look at a map of Earth as a flat, two-dimensional representation to know that shape has everything to do with spatial interpretation. Earth distances estimated on a flat map are much different from the same distances measured on a globe. Figure **22.2** shows interpretations of the shape of the universe. Consider the cosmic implications of a concept as simple as the wheel, especially when given its universal importance in ancient religions. Does the center of the wheel turn? How does the space-time interval at the edge of the wheel relate to that of the center? Our perception of reality is very dependent on our point of view.

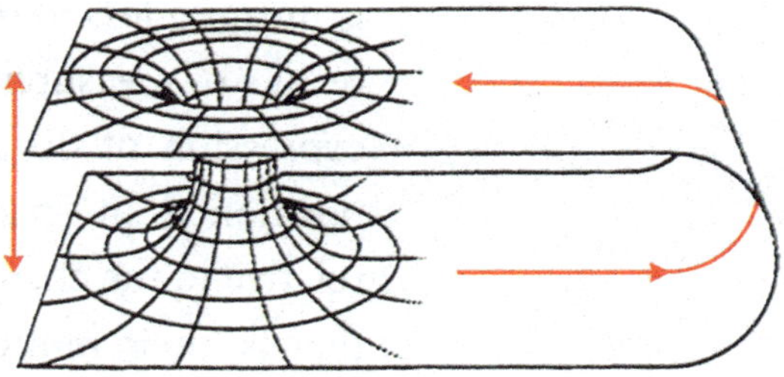

22.2. Einsteinian space wormhole. Light traveling near an object having a strong gravitational force bends to conform to the space around the object. Instead of traveling in a straight line, it will follow the line traced out by one of the isometric gravitational lines. W. J. Kaufmann, ***Relativity and Cosmology*.**

233

In a universe populated by islands of galaxies and black holes with massive gravitational presences, the rhythmic flow of time and space, recorded in the finite messages of light and energy reaching Earth, relate more to the spiritual force of the Algonquin Manitou than to the mechanical laws of absolute Newtonian physics. To this end, Einstein's relativity theories are the deepest penetration into space-time reality ever attempted by humankind. Einstein acknowledged his trespass on theology when he said in 1931, "The cosmic religious experience is the strongest and the noblest driving force behind scientific research." He went even further in 1940 to say, "Science without religion is lame, religion without science is blind."

The German physicist Werner Heisenberg (who won a Nobel Prize in 1932 for work he carried out when only 24) pointed out that it is meaningless to claim measurement on an atomic scale, because the mere act of taking a measurement will alter the system being measured. The Heisenberg Uncertainty Principle, or Indeterminacy Principle, declares that an exact measurement is impossible because the act of measuring changes the event measured, at least in the case of atomic particles. Thus, change in the event measured is unaccountable. Implicit in the Heisenberg Principle is the inability to predict moment-to-moment existence of a system. A macrocosmic correlate of the Heisenberg Principle might say, in the tradition of Zeno, that we cannot observe position and motion at the same time. It is meaningless to describe position or motion except as results of measurements. Therefore, the light year as an expression of distance is meaningless because objects positioned in space define space. We measure time relative to the change of the position of objects in space, that is, their motions. Therefore, time *and* space have meaning only if something is moving relative to something else. In the universe, it is not at all clear if motion and position are a function of time and space or if time and space are a function of cosmic principles that we do not yet understand.

Immanuel Kant concluded that space and time are not real; they are inventions of our perception. Like standing at the center of a wheel whose spokes reach out to infinity; or watching a jet liner slowly traverse the celestial dome in space — time intervals that defy our intellectual expectations of jet travel; or seeing a planet in the evening sky with no perception of the incomparable velocity at which it is moving. Objects do not impose space and time on our minds, wrote Kant; our minds impose space and time on objects; it is our weakness for finite reasoning. Comprehension of the infinite is beyond human capability. It is therefore impossible to know the universe. To try to do so is to fall into Zeno's trap of irreconcilable contradictions. The only way out is to change the system by which time and space are ultimately perceived and measured.

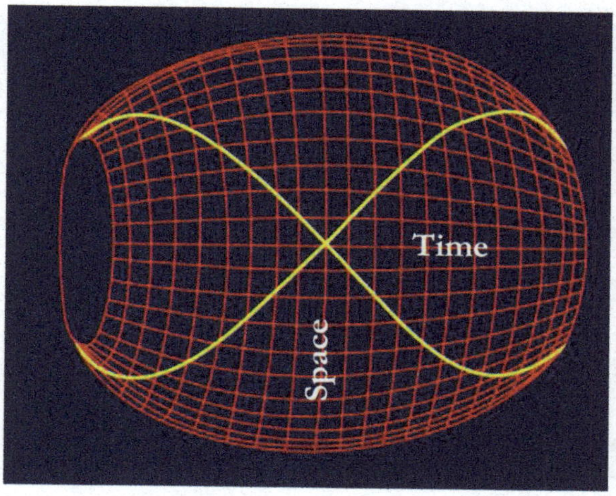

22.3. Curved space-time. The time coordinate of one coordinate system depends on both the time and space coordinates of another relatively moving system. ***Britannica*.**

234

Is the universe finite or infinite, stable or moving, closed or open? What is its shape? Scientists working at the end of the 20th century used leftover radiation from the Big Bang as a microwave map to show that the universe is flat. Uninformed explorers of the ancient world once believed the same about Earth. The human mind cannot conceive any description that would satisfactorily explain in its entirety something as vast as the universe that goes on potentially forever, or know what, if anything might lie beyond its edge. The question of being leaves science to contend with a few comprehensible parts of an incomprehensible whole. The universe is a theoretical process, revealing its great secrets little by little and ever so slowly, and then only to the most enlightened thinkers.

INDEX

Made in United States
Orlando, FL
04 November 2026

90206683R00221